Manual for Streets

Published by Thomas Telford Publishing, Thomas Telford Ltd, 1 Heron Quay, London E14 4JD. www.thomastelford.com

Distributors for Thomas Telford books are
USA: ASCE Press, 1801 Alexander Bell Drive, Reston, VA 20191-4400, USA
Japan: Maruzen Co. Ltd, Book Department, 3–10 Nihonbashi 2-chome, Chuo-ku, Tokyo 103
Australia: DA Books and Journals, 648 Whitehorse Road, Mitcham 3132, Victoria

First published 2007

Published for the Department for Transport under licence from the Controller of Her Majesty's Stationery Office

A catalogue record for this book is available from the British Library

ISBN: 978-0-7277-3501-0

recycle

Printed and bound in Great Britain by Maurice Payne Colourprint Limited using material containing at least 75% recycled fibre.

Ordnance Survey mapping
All mapping is reproduced from Ordnance Survey material with the permission of Ordnance Survey on behalf of the Controller of Her Majesty's Stationery Office © Crown copyright. Unauthorised reproduction infringes Crown copyright and may lead to prosecution or civil proceedings. Department for Transport 100039241, 2007.

Cover image © Countryside Properties. Scheme designed by MDA

Contents

Acknowledgements

Project team
Manual for Streets was produced by a team led by consultants WSP, with Llewelyn Davies Yeang (LDY), Phil Jones Associates (PJA) and TRL Limited on behalf of the Department for Transport, and Communities and Local Government.

The core team comprised (all lists in alphabetical order):
· Annabel Bradbury (TRL)
· Andrew Cameron (WSP)
· Ben Castell (LDY)
· Phil Jones (PJA)
· Tim Pharoah (LDY),
· Stuart Reid (TRL)
· Alan Young – Project Manager, (WSP)

With additional research and assistance by:
Sam Carman (WSP), Tom Ewings (TRL), Una McGaughrin (LDY) Peter O'Brien (LDY), Ross Paradise (TRL), Christianne Strubbe (Hampshire County Council), Iain York (TRL)

Graphic design by Llewelyn Davies Yeang (Ros Shakibi, Ting LamTang and Thanh Tung Uong, with artwork by Alexandra Steed) and overseen by Ela Ginalska (Department for Transport)

Steering group
The Project Steering Group included:
Bob Bennett (Planning Officers Society), Edward Chorlton (Devon County Council), Vince Christie (Local Government Association), Wayne Duerden (Department for Transport) Louise Duggan (Commission for Architecture and the Built Environment), Ray Farrow (Home Builders' Federation) George Hazel (Urban Design Alliance), Ed Hobson (Commission for Architecture and the Built Environment), Gereint Killa (Department for Transport), Grahame Lawson (Disabled Persons Transport Advisory Committee), Spencer Palmer (Department for Transport), John Smart (Institution of Highways and Transportation), Larry Townsend (Communities and Local Government), Polly Turton (Commission for Architecture and the Built Environment), David Williams (Department for Transport), Mario Wolf (Communities and Local Government), Philip Wright (Health & Safety Executive)

Sounding board
Further advice was received from an invited Sounding Board consisting of:
Tony Aston (Guide Dogs for the Blind Association), David Balcombe (Essex County Council), Peter Barker (Guide Dogs for the Blind Association), Richard Button (Colchester Borough Council) Jo Cleary (Friends of the Lake District), Meredith Evans (Borough of Telford & Wrekin Council), Tom Franklin (Living Streets), Jenny Frew (English Heritage), Stephen Hardy (Dorset County Council), Richard Hebditch (Living Streets), Ian Howes (Colchester Borough Council), Andrew Linfoot (Halcrow), Peter Lipman (Sustrans), Ciaran McKeon (Dublin Transport Office), Elizabeth Moon, (Essex County Council), Nelia Parmaklieva (Colchester Borough Council), Mark Sackett (RPS), Paul Sheard (Leicestershire County Council), Alex Sully (Cycling England), Carol Thomas (Guide Dogs for the Blind Association), Andy Yeomanson (Leicestershire County Council), Emily Walsh (Solihull Metropolitan Borough Council), Leon Yates (London Borough of Lewisham)

Additional consultation and advice
Additional consultation took place with the following:
Mark Ainsworth (George Wimpey), John Barrell (Jacobs Consultancy), Terry Brown (GMW Architects), Hywel Butts (Welsh Assembly Government), David Coatham (Institution of Lighting Engineers), Mike Darwin (Leeds City Council), Adrian Lord (Arup / Cycling England), Kevin Pearson (Avon Fire & Rescue Service), Michael Powis (Nottinghamshire Police), Gary Kemp (Disabled Persons Transport Advisory Committee), Malcolm Lister (London Borough of Hounslow)

In addition to those already listed, substantial comments on drafts of the manual were received from:
Duncan Barratt (West Sussex County Council), Neil Benison (Warwickshire County Council), Daniel Black (Sustrans), Rob Carmen (Medway Council), Greg Devine (Surrey County Council), John Emslie (MVA Consultancy), Heather Evans (Cyclists' Touring Club), David Groves (Cornwall County Council), Steve Mead (Derbyshire County Council), Christine Robinson (Essex County Council), Mick Sankus (Medway Council), Mike Schneider (North Somerset Borough Council), Graham Paul Smith (Oxford Brookes University), Fiona Webb (Mid Bedfordshire District Council), Bob White (Kent County Council)

Case studies
A number of case studies were investigated in the compilation if the Manual. These are listed below, along with the individuals who provided assistance:
· Beaulieu Park, Chelmsford:
  Sarah Hill-Sanders, Chelmsford Borough Council
  Chris Robinson, Essex County Council
· Charlton Down, Dorset:
  Stephen Hardy, Dorset County Council
  Ian Madgwick, Dorset County Council
· Crown Street, Glasgow:
  Elaine Murray, Glasgow City Council
  Mic Ralph, Glasgow City Council
  Stephen Rigg, CZWG Architects
· Darwin Park, Lichfield:
  Steve Clarke, Staffordshire County Council
  Ian Thompson, Lichfield District Council
· Hulme, Manchester:
  Kevin Gillham, Manchester City Council
  Brian Kerridge, Manchester City Council
· Limehouse Fields, Tower Hamlets:
  Angelina Eke, Tower Hamlets Borough Council
  John Hilder, Tower Hamlets Borough Council
· New Hall, Harlow:
  Alex Cochrane, Roger Evans Associates
  Keith Lawson, Essex County Council
  Mriganka Saxena, Roger Evans Associates
· Pirelli site, Eastleigh:
  Dave Francis, Eastleigh Borough Council
  Eric Reed, Eastleigh Borough Council
· Queen Elizabeth Park, Guildford:
  David Barton, Guildford Borough Council
  David Taylor, Surrey County Council
· Staithes South Bank, Gateshead:
  Alastair Andrew, Gateshead Council
  Andy Szandrowski, Gateshead Council

Status and application

Manual for Streets (MfS) supersedes Design Bulletin 32 and its companion guide Places, Streets and Movement, which are now withdrawn in England and Wales. It complements Planning Policy Statement 3: Housing and Planning Policy Wales. MfS comprises technical guidance and does not set out any new policy or legal requirements.

MfS focuses on lightly-trafficked residential streets, but many of its key principles may be applicable to other types of street, for example high streets and lightly-trafficked lanes in rural areas. It is the responsibility of users of MfS to ensure that its application to the design of streets not specifically covered is appropriate.

MfS does not apply to the trunk road network. The design requirements for trunk roads are set out in the Design Manual for Roads and Bridges (DMRB).

MfS only applies formally in England and Wales.

The policy, legal and technical frameworks are generally the same in England and Wales, but where differences exist these are made clear.

Foreword

Streets are the arteries of our communities – a community's success can depend on how well it is connected to local services and the wider world. However, it is all too easy to forget that streets are not just there to get people from A to B. In reality, streets have many other functions. They form vital components of residential areas and greatly affect the overall quality of life for local people.

Places and streets that have stood the test of time are those where traffic and other activities have been integrated successfully, and where buildings and spaces, and the needs of people, not just of their vehicles, shape the area. Experience suggests that many of the street patterns built today will last for hundreds of years. We owe it to present and future generations to create well-designed places that will serve the needs of the local community well.

In 2003, we published detailed research[1] which demonstrated that the combined effect of the existing policy, legal and technical framework was not helping to generate consistently good quality streets. Without changes this framework was holding back the creation of the sustainable residential environments that communities need and deserve.

As a society, we have learned to appreciate the value of a clear and well-connected street network, well defined public and private spaces, and streets that can be used in safety by a wide range of people. We also understand the benefits of ensuring that the different functions of streets are integral to their design from the outset. But we need to do more to recognise the role that streets play in the life of a community, particularly the positive opportunities that they can bring for social interaction. To achieve this we need strong leadership and clear vision. Importantly, we need to tackle climate change, and helping and encouraging people to choose more sustainable ways of getting around will be key.

Manual for Streets explains how to respond to these issues. Although it does not set out new policy or legislation, it shows how the design of residential streets can be enhanced. It also advises on how street design can help create better places – places with local distinctiveness and identity. In addition, it establishes a common reference point for all those involved in the design of residential neighbourhoods.

This publication represents a strong Government and Welsh Assembly commitment to the creation of sustainable and inclusive public spaces. We hope that everyone who plays a part in making and shaping the built environment will embrace its principles to help deliver places that work for communities now, and in the future.

1 DfT, ODPM (July 2003) *Better Streets, Better Places – Delivering Sustainable Residential Environments: PPG3 and Highway Adoption*. London: ODPM.

Gillian Merron MP
Transport Minister

Baroness Andrews OBE
Parliamentary Under Secretary of State Communities and Local Government

Tamsin Dunwoody AM
Deputy Minister for Enterprise, Innovation and Networks Deputy Minister for Environment, Planning & Countryside

Preface

Manual for Streets (MfS) replaces Design Bulletin 32, first published in 1977, and its companion guide Places, Streets and Movement. It puts well-designed residential streets at the heart of sustainable communities.

For too long the focus has been on the movement function of residential streets. The result has often been places that are dominated by motor vehicles to the extent that they fail to make a positive contribution to the quality of life. MfS demonstrates the benefits that flow from good design and assigns a higher priority to pedestrians and cyclists, setting out an approach to residential streets that recognises their role in creating places that work for all members of the community. MfS refocuses on the place function of residential streets, giving clear guidance on how to achieve well-designed streets and spaces that serve the community in a range of ways.

MfS updates the link between planning policy and residential street design. It challenges some established working practices and standards that are failing to produce good-quality outcomes, and asks professionals to think differently about their role in creating successful neighbourhoods. It places particular emphasis on the importance of collaborative working and coordinated decision-making, as well as on the value of strong leadership and a clear vision of design quality at the local level.

Research carried out in the preparation of Manual for Streets indicated that many of the criteria routinely applied in street design are based on questionable or outdated practice. For example, it showed that, when long forward visibility is provided and generous carriageway width is specified, driving speeds tend to increase. This demonstrates that driver behaviour is not fixed; rather, it can be influenced by the environment. MfS addresses these points, recommending revised key geometric design criteria to allow streets to be designed as places in their own right while still ensuring that road safety is maintained.

MfS is clear that uncoordinated decision-making can result in disconnected, bland places that fail to make a contribution to the creation of thriving communities. It recommends that development teams are established to negotiate issues in the round and retain a focus on the creation of locally distinct, high-quality places. Where high levels of change are anticipated, designers and other stakeholders are encouraged to work together strategically from an early stage. MfS also recommends the use of tools such as masterplans and design codes.

Neighbourhoods where buildings, streets and spaces combine to create locally distinct places and which make a positive contribution to the life of local communities need to become more widespread. MfS provides a clear framework for the use of local systems and procedures; it also identifies the tools available to ensure that growth and change are planned for and managed in an integrated way. The aspirations of MfS – interdisciplinary working, strategic coordination and balanced decision making – will only become a reality if they are developed and applied at a local level. This is already happening in some places, and the results are promising – this document aims to make the adoption of such practice the norm.

MfS does not set out new policy or introduce new additional burdens on local authorities, highway authorities or developers. Rather it presents guidance on how to do things differently within the existing policy, technical and legal framework.

A

Context and process

1

Introduction

Chapter aims

- Set out the aims of *Manual for Streets*.

- Explain the status of *Manual for Streets* and its relationship with local design standards and the *Design Manual for Roads and Bridges*.

- Promote greater collaboration between all those involved in the design, approval and adoption processes.

- Summarise key changes from previous guidance.

Figure 1.1 Streets should be attractive places that meet the needs of all users.

1.1   Aims of the document

1.1.1   There is a need to bring about a transformation in the quality of streets. This requires a fundamental culture change in the way streets are designed and adopted, including a more collaborative approach between the design professions and other stakeholders. People need to think creatively about their various roles in the process of delivering streets, breaking away from standardised, prescriptive, risk-averse methods to create high-quality places.

1.1.2   Streets make up the greater part of the public realm. Better-designed streets therefore contribute significantly to the quality of the built environment and play a key role in the creation of sustainable, inclusive, mixed communities consistent with the policy objectives of *Planning Policy Statement 1: Delivering Sustainable Development* (PPS1)[1], *Planning Policy Statement 3: Housing* (PPS3)[2] and *Planning Policy Wales* (PPW).[3]

1.1.3   *Manual for Streets* (MfS) is expected to be used predominantly for the design, construction, adoption and maintenance of new residential streets, but it is also applicable to existing residential streets subject to re-design. For new streets, MfS advocates a return to more traditional patterns which are easier to assimilate into existing built-up areas and which have been proven to stand the test of time in many ways.

1.1.4   Streets should not be designed just to accommodate the movement of motor vehicles. It is important that designers place a high priority

on meeting the needs of pedestrians, cyclists and public transport users, so that growth in these modes of travel is encouraged (Fig. 1.1).

1.1.5   MfS aims to assist in the creation of streets that:
- help to build and strengthen the communities they serve;
- meet the needs of all users, by embodying the principles of inclusive design (see box);
- form part of a well-connected network;
- are attractive and have their own distinctive identity;
- are cost-effective to construct and maintain; and
- are safe.

The principles of inclusive design

Inclusive design:[4]
- places people at the heart of the design process;
- acknowledges diversity and difference;
- offers choice where a single solution cannot accommodate all users;
- provides for flexibility in use; and
- provides buildings and environments that are convenient and enjoyable to use for everyone.

1.1.6   MfS discourages the building of streets that are:
- primarily designed to meet the needs of motor traffic;
- bland and unattractive;
- unsafe and unwelcoming to pedestrians and cyclists;
- difficult to serve by public transport; and
- poorly designed and constructed (Fig. 1.2).

1   Office of the Deputy Prime Minister (ODPM) (2005) *Planning Policy Statement 1: Delivering Sustainable Development.* London: The Stationery Office (TSO).
2   Communities and Local Government (2006) *Planning Policy Statement 3: Housing.* London: TSO.
3   Welsh Assembly Government (2002). *Planning Policy Wales.* Cardiff: National Assembly for Wales (NAfW). Chapter 2, Planning for Sustainability.
4   Commission for Architecture and the Built Environment (CABE) (2006) *The Principles of Inclusive Design (They Include You).* London: CABE. (*Wales*: See also Welsh Assembly Government (2002). *Technical Advice Note 12: Design.* Cardiff: NAfW. Chapter 5, Design Issues.)

1.1.7   For the purposes of this document, *a street is defined as a highway that has important public realm functions beyond the movement of traffic*. Most critically, streets should have a sense of place, which is mainly realised through local distinctiveness and sensitivity in design. They also provide direct access to the buildings and the spaces that line them. Most highways in built-up areas can therefore be considered as streets.

1.2   Who the manual is for

1.2.1   MfS is directed to all those with a part to play in the planning, design, approval or adoption of new residential streets, and modifications to existing residential streets. This includes the following (in alphabetical order):

- Organisations:
  - developers;
  - disability and other user groups;
  - emergency services;
  - highway and traffic authorities;
  - planning authorities;
  - public transport providers;
  - utility and drainage companies; and
  - waste collection authorities.
- Professions:
  - access/accessibility officers;
  - arboriculturists;
  - architects;
  - drainage engineers;
  - highway/traffic engineers;
  - landscape architects;
  - local authority risk managers;
  - police architectural liaison officers and crime prevention officers;
  - road safety auditors;
  - street lighting engineers;
  - town planners;
  - transport planners;
  - urban designers.

1.2.2   These lists are not exhaustive and there are other groups with a stake in the design of streets. Local communities, elected members and civic groups, in particular, are encouraged to make use of this document.

1.2.3   *MfS covers a broad range of issues and it is recommended that practitioners read every section regardless of their specific area of interest.* This will create a better understanding of the many and, in some cases, conflicting

Figure 1.2 Streets should not be bland and unwelcoming.

priorities that can arise. A good design will represent a balance of views with any conflicts resolved through compromise and creativity.

1.3   Promoting joint working

1.3.1   In the past street design has been dominated by some stakeholders at the expense of others, often resulting in unimaginatively designed streets which tend to favour motorists over other users.

1.3.2   MfS aims to address this by encouraging a more holistic approach to street design, while assigning a higher priority to the needs of pedestrians, cyclists and public transport. The intention is to create streets that encourage greater social interaction and enjoyment while still performing successfully as conduits for movement.

1.3.3   It is important for the various parts of local government to work together when giving input to a development proposal. Developers may be faced with conflicting requirements if different parts of local government fail to coordinate their input. This can cause delay and a loss of design quality. This is particularly problematic when one section of a local authority – for example the highway adoption or maintenance engineers – become involved late on in the process and require significant changes to the design. A collaborative process is required from the outset.

1.4   DMRB and other design standards

1.4.1   The Department for Transport does not set design standards for highways – these are set by the relevant highway authority.

1.4.2   The Secretary of State for Transport is the highway authority for trunk roads in England and acts through the Highways Agency (HA). In Wales the Welsh Assembly Government is the highway authority for trunk roads. The standard for trunk roads is the *Design Manual for Roads and Bridges* (DMRB).[5]

1.4.3   Some trunk roads could be described as 'streets' within the definition given in MfS, but their strategic nature means that traffic movement is their primary function. MfS does not apply to trunk roads.

1.4.4   The DMRB is not an appropriate design standard for most streets, particularly those in lightly-trafficked residential and mixed-use areas.

1.4.5   Although MfS provides guidance on technical matters, local standards and design guidance are important tools for designing in accordance with the local context. Many local highway authorities have developed their own standards and guidance. Some of these documents, particularly those published in recent years, have addressed issues of placemaking and urban design, but most have not. *It is therefore strongly recommended that local authorities review their standards and guidance to embrace the principles of MfS.* Local standards and guidance should focus on creating and improving local distinctiveness through the appropriate choice of layouts and materials while adhering to the overall guidance given in MfS.

1.5   Development of Manual for Streets

1.5.1   The preparation of MfS was recommended in *Better Streets, Better Places,*[6] which advised on how to overcome barriers to the creation of better quality streets.

1.5.2   MfS has been produced as a collaborative effort involving a wide range of key stakeholders with an interest in street design. It has been developed by a multi-disciplinary team of highway engineers, urban designers, planners and researchers. The recommendations contained herein are based on a combination of:
- primary research;
- a review of existing research;
- case studies;

- existing good practice guidance; and
- consultation with stakeholders and practitioners.

1.5.1   During its preparation, efforts have been made to ensure that MfS represents a broad consensus and that it is widely accepted as good practice.

1.6   Changes in approach

1.6.1   The main changes in the approach to street design that MfS recommends are as follows:

- applying a user hierarchy to the design process with pedestrians at the top;
- emphasising a collaborative approach to the delivery of streets;
- recognising the importance of the community function of streets as spaces for social interaction;
- promoting an inclusive environment that recognises the needs of people of all ages and abilities;
- reflecting and supporting pedestrian desire lines in networks and detailed designs;
- developing masterplans and preparing design codes that implement them for larger-scale developments, and using design and access statements for all scales of development;
- creating networks of streets that provide permeability and connectivity to main destinations and a choice of routes;
- moving away from hierarchies of standard road types based on traffic flows and/or the number of buildings served;
- developing street character types on a location-specific basis with reference to both the place and movement functions for each street;
- encouraging innovation with a flexible approach to street layouts and the use of locally distinctive, durable and maintainable materials and street furniture;
- using quality audit systems that demonstrate how designs will meet key objectives for the local environment;
- designing to keep vehicle speeds at or below 20 mph on residential streets unless there are overriding reasons for accepting higher speeds; and
- using the minimum of highway design features necessary to make the streets work properly.

5   Highways Agency (1992) *Design Manual for Roads and Bridges*. London: TSO.
6   ODPM and Department for Transport (2003) *Better Streets, Better Places: Delivering Sustainable Residential Environments*; PPG3 and Highway Adoption London: TSO.

2

Streets in context

Chapter aims

- Explain the distinction between 'streets' and 'roads'.

- Summarise the key functions of streets.

- Propose a new approach to defining street hierarchies, based on their significance in terms of both place and movement.

- Set out the framework of legislation, standards and guidance that apply to the design of streets.

- Provide guidance to highway authorities in managing their risk and liability.

2.1 Introduction

2.1.1 This chapter sets out the overall framework in which streets are designed, built and maintained.

2.1.2 *The key recommendation is that increased consideration should be given to the 'place' function of streets.* This approach to addressing the classification of streets needs to be considered across built-up areas, including rural towns and villages, so that a better balance between different functions and street users is achieved.

2.2 Streets and roads

2.2.1 A clear distinction can be drawn between streets and roads. Roads are essentially highways whose main function is accommodating the movement of motor traffic. Streets are typically lined with buildings and public spaces, and while movement is still a key function, there are several others, of which the place function is the most important (see 'Streets – an historical perspective' box).

Streets – an historical perspective

Most historic places owe their layout to their original function. Towns have grown up around a market place (Fig. 2.1), a bridgehead or a harbour; villages were formed according to the pattern of farming and the ownership of the land. The layouts catered mostly for movement on foot. The era of motorised transport and especially privately-owned motor vehicles has, superficially at least, removed the constraint that kept urban settlements compact and walkable.

When the regulation of roads and streets began, spread of fire was the main concern. Subsequently health came to the forefront and the classic 36 ft wide bye-law street was devised as a means of ensuring the passage of air in densely built-up areas. Later, the desire to guarantee that sunshine would get to every house led to the requirement for a 70 ft separation between house fronts, and this shaped many developments from the 1920s onwards.

It was not until after the Second World War, and particularly with the dramatic increase in car ownership from the 1960s onwards, that traffic considerations came to dominate road design.

Figure 2.1 Newark: (a) the Market Place, 1774; and (b) in 2006.

Andrew Cameron

2.2.2   Streets have to fulfil a complex variety of functions in order to meet people's needs as places for living, working and moving around in. This requires a careful and multi-disciplinary approach that balances potential conflicts between different objectives.

2.2.3   In the decades following the Second World War, there was a desire to achieve a clear distinction between two types of highway:
- distributor roads, designed for movement, where pedestrians were excluded or, at best, marginalised; and
- access roads, designed to serve buildings, where pedestrians were accommodated.

This led to layouts where buildings were set in the space between streets rather than on them, and where movement on foot and by vehicle was segregated, sometimes using decks, bridges or subways. Many developments constructed using such layouts have had significant social problems and have either been demolished or undergone major regeneration (Fig. 2.2).

2.2.4   This approach to network planning limited multi-functional streets to the most lightly-trafficked routes. This has led to development patterns where busy distributor roads link relatively small cells of housing. Such layouts are often not conducive to anything but the shortest of trips on foot or by bicycle. It is now widely recognised that there are many advantages in extending the use of multi-functional streets in urban areas to busier routes.

2.2.5   Streets that are good quality places achieve a number of positive outcomes, creating a virtuous circle:
- attractive and well-connected permeable street networks encourage more people to walk and cycle to local destinations, improving their health while reducing motor traffic, energy use and pollution;[1]
- more people on the streets leads to improved personal security and road safety – research shows that the presence of pedestrians on streets causes drivers to travel more slowly;[2]
- people meeting one another on a casual basis strengthens communities and encourages a sense of pride in local environments; and
- people who live in good-quality environments are more likely to have a sense of ownership and a stake in maintaining the quality of their local streets and public spaces.

2.2.6   Well-designed streets thus have a crucial part to play in the delivery of sustainable communities, defined as 'places where people want to live and work, now and in the future'.[3]

2.2.7   Lanes in rural areas can provide other functions than just movement, including various leisure activities such as walking, cycling and horse riding.

1   Snellen, D. (1999) The relationship between urban form and activity patterns. In *Proceedings of the European Transport Conference, Cambridge*, 1999. London: PTRC. pp. 429–439.
2   ODPM and Home Office (2004) *Safer Places: The Planning System and Crime Prevention*. London: TSO.
3   ODPM (2005) *Planning Policy Statement 1: Delivering Sustainable Developments*. London: TSO. (*Wales*: Welsh Assembly Government (2002) *Planning Policy Wales*. Cardiff: NAfW.)

Figure 2.2 A poor-quality space with a layout where pedestrians and vehicles are segregated. It has not been a success and the area is now undergoing regeneration.

Manual for Streets

2.3   Principal functions of streets

2.3.1   Streets have five principal functions;
- place;
- movement;
- access;
- parking; and
- drainage, utilities and street lighting.

These functions are derived from *Paving the Way*.[4]

Place

2.3.2   The place function is essentially what distinguishes a street from a road. The sense of place is fundamental to a richer and more fulfilling environment. It comes largely from creating a strong relationship between the street and the buildings and spaces that frame it. The Local Government White Paper[5] makes it clear that, in creating sustainable communities, local authorities have an essential and strategic role.

2.3.3   An important principle was established in *Places, Streets and Movement*[6] – *when planning new developments, achieving a good place should come before designing street alignments, cross-sections and other details.* Streets should be fitted around significant buildings, public spaces, important views, topography, sunlight and microclimate.

2.3.4   A sense of place encompasses a number of aspects, most notably the street's:
- local distinctiveness;
- visual quality; and
- propensity to encourage social activity (Fig. 2.3).

These are covered in more detail in Chapters 4 and 5.

2.3.5   The choice of surface materials, planting and street furniture has a large part to play in achieving a sense of place. The excessive or insensitive use of traffic signs and other street furniture has a negative impact on the success of the street as a place. It is particularly desirable to minimise the environmental impact of highway infrastructure in rural areas, for example, where excessive lighting and the inappropriate use of kerbing, signs, road markings and street furniture can urbanise the environment.

Movement

2.3.6   Providing for movement along a street is vital, but it should not be considered independently of the street's other functions. The need to cater for motor vehicles is well understood by transport planners, but the passage of people on foot and cycle has often been neglected. Walking and cycling are important modes of travel, offering a more sustainable alternative to the car, making a positive contribution to the overall character of a place, public health and to tackling climate change through reductions in carbon emissions. Providing for movement is covered in more detail in Chapters 6 and 7.

4   Commission for Architecture and the Built Environment (CABE) and ODPM (2002) *Paving the Way: How we Achieve Clean, Safe and Attractive Streets*. London: Thomas Telford Ltd.
5   Communities and Local Government (2006) *Strong and Prosperous Communities: The Local Government White Paper*. London: TSO.
6   Department for Environment, Transport and the Regions (DETR) (1998) *Places, Streets and Movement: A Companion Guide to Design Bulletin 32 – Residential Roads and Footpaths*. London: TSO.

Lorraine Farrelly

Figure 2.3 A residential environment showing distinctive character.

Llewelyn Davies Yeang

Figure 2.4 An example of a sustainable drainage system.

Access

2.3.7   Access to buildings and public spaces is another important function of streets. Pedestrian access should be designed for people of all ages and abilities.

2.3.8   Providing frontages that are directly accessible on foot and that are overlooked from the street is highly desirable in most circumstances as this helps to ensure that streets are lively and active places. The access function is covered in Chapters 6 and 7.

Parking

2.3.9   Parking is a key function of many streets, although it is not always a requirement. A well-designed arrangement of on-street parking provides convenient access to frontages and can add to the vitality of a street. Conversely, poorly designed parking can create safety problems and reduce the visual quality of a street. Parking is covered in more detail in Chapter 8.

Drainage, utilities and street lighting

2.3.10   Streets are the main conduits for drainage and utilities. Buried services can have a major impact on the design and maintenance requirements of streets. Sustainable drainage systems can bring environmental benefits, such as flood control, creating wildlife habitats and efficient wastewater recycling (Fig. 2.4). Drainage and utilities are covered in Chapter 11, and street lighting is covered in Chapter 10.

2.4   The balance between place and movement

2.4.1   Of the five functions, place and movement are the most important in determining the character of streets.

2.4.2   In the past, road design hierarchies have been based almost exclusively on the importance attributed to vehicular movement. This has led to the marginalisation of pedestrians and cyclists in the upper tiers where vehicular capacity requirements predominate. The principle that a road was primarily for motor traffic has tended to filter down into the design of streets in the bottom tiers of the hierarchy.

2.4.3   This approach has created disjointed patterns of development. High-speed roads often have poor provision for pedestrian activity, cutting residential areas off from each other and from other parts of a settlement. In addition, the hierarchy does not allow for busy arterial streets, which feature in most towns and cities.

2.4.4   Streets should no longer be designed by assuming 'place' to be automatically subservient to 'movement'. Both should be considered in combination, with their relative importance depending on the street's function within a network. It is only by considering both aspects that the right balance will be achieved. It is seldom appropriate to focus solely on one to the exclusion of the other, even in streets carrying heavier volumes of traffic, such as high streets.

2.4.5   Place status denotes the relative significance of a street, junction or section of a street in human terms. The most important places will usually be near the centre of any settlement or built-up area, but important places will also exist along arterial routes, in district centres, local centres and within neighbourhoods.

2.4.6   Movement status can be expressed in terms of traffic volume and the importance of the street, or section of street, within a network – either for general traffic or within a mode-specific (e.g. bus or cycle) network. It can vary along the length of a route, such as where a street passes through a town centre.

Movement status

Motorway

High street

Residential street

Place status

Figure 2.5 Typical road and street types in the Place and Movement hierarchy.

2.4.7   Highway authorities assess the relative importance of particular routes within an urban area as part of their normal responsibilities, such as those under the New Roads and Streetworks Act 1991.[7] One of the network management duties under the Traffic Management Act 2004[8] is that all local traffic authorities should determine specific policies or objectives for different roads or classes of road in their road network. See also the *Network Management Duties Guidance*[9] published by the Department for Transport in November 2004 (*Wales*: guidance published November 2006[10]). This states that it is for the authority to decide the levels of priority given to different road users on each road, for example, particular routes may be defined as being important to the response times of the emergency services.

2.4.8   Another way of assessing the movement status of a street is to consider the geographical scale of the destinations it serves. Here, movement status can range from national networks (including motorways) through to city, town, district, neighbourhood and local networks, where the movement function of motor vehicles would be minimal.

Place and movement matrix

2.4.9   Defining the relative importance of particular streets/roads in terms of place and movement functions should inform subsequent design choices. For example:

- motorways – high movement function, low place function;
- high streets – medium movement function, medium to high place function; and
- Residential streets – low to medium movement function, low to medium place function.

2.4.10   This way of looking at streets can be expressed as a two-dimensional hierarchy,[11] where the axes are defined in terms of place and movement (Fig. 2.5). It recognises that, whilst some streets are more important than others in terms of traffic flow, some are also more important than others in terms of their place function and deserve to be treated differently. This approach allows designers to break away from previous approaches to hierarchy, whereby street designs were only based on traffic considerations.

7   New Roads and Street Works Act 1991. London: TSO.
8   Traffic Management Act 2004. London: TSO.
9   Department for Transport (2004) *Network Management Duties Guidance*. London: TSO.
10 Welsh Assembly Government (2006) *Traffic Management Act 2004 Network Management Duty Guidance*. Cardiff: NAfW.
11 The two-dimensional hierarchy as a way of informing street design was developed by the EU project ARTISTS. See www.tft.lth.se/artists/

2.4.11   In many situations it will be possible to determine the place status of existing streets by consulting with the people living there. Such community consultation is encouraged.

2.4.12   In new developments, locations with a relatively high place function would be those where people are likely to gather and interact with each other, such as outside schools, in local town and district centres or near parades of shops. Streets that pass through these areas need to reflect these aspects of their design, which will have been identified at the masterplan/scheme design stage.

2.4.13   Once the relative significance of the movement and place functions has been established, it is possible to set objectives for particular parts of a network. This will allow the local authority to select appropriate design criteria for creating new links or for changing existing ones.

2.4.14   Movement and place considerations are important in determining the appropriate design speeds, speed limits and road geometry, etc., along with the level of adjacent development and traffic composition (see Department for Transport Circular 01/2006;[12] *Wales*: Welsh Office Circular 1/1993[13]).

2.5   Policy, legal and technical context

2.5.1   There is a complex set of legislation, polices and guidance applying to the design of highways. There is a tendency among some designers to treat guidance as hard and fast rules because of the mistaken assumption that to do otherwise would be illegal or counter to a stringent policy. This tends to restrict innovation, leading to standardised streets with little sense of place or quality. In fact, there is considerable scope for designers and approving authorities to adopt a more flexible approach on many issues.

2.5.2   The following comprise the various tiers of instruction and advice:
· the legal framework of statutes, regulations and case law;
· government policy;
· government guidance;
· local policies;
· local guidance; and
· design standards.

2.5.3   Parliament and the courts establish the legal framework within which highway authorities, planning authorities and other organisations operate.

2.5.4   The Government develops policies aimed at meeting various objectives which local authorities are asked to follow. It also issues supporting guidance to help authorities implement these policies.

2.5.5   Within this overall framework highway and planning authorities have considerable leeway to develop local policies and standards, and to make technical judgements with regard to how they are applied. Other bodies also produce advisory and research material that they can draw on.

2.6   Risk and liability

2.6.1   A major concern expressed by some highway authorities when considering more innovative designs, or designs that are at variance with established practice, is whether they would incur a liability in the event of damage or injury.

2.6.2   This can lead to an over-cautious approach, where designers strictly comply with guidance regardless of its suitability, and to the detriment of innovation. This is not conducive to creating distinctive places that help to support thriving communities.

2.6.3   In fact, imaginative and context-specific design that does not rely on conventional standards can achieve high levels of safety. The design of Poundbury in Dorset, for example, did not comply fully with standards and guidance then extant, yet it has few reported accidents. This issue was explored in some detail in the publication *Highway Risk and Liability Claims*.[14]

2.6.4   Most claims against highway authorities relate to alleged deficiencies in maintenance. The duty of the highway authority to maintain the highway is set out in section 41 of the Highways Act 1980,[15] and case law has clarified the law in this area.

12 Department for Transport (2006) *Setting Local Speed Limits*. Circular 01/2006. London: TSO
13 Department for Transport and Welsh Office (1993) *Welsh Office Circular 01/1993. Road Traffic Regulation Act 1984: Sections 81–85 Local Speed Limits*. Cardiff: Welsh Office.
14 UK Roads Board (2005) *Highway Risk and Liability Claims – A Practical Guide to Appendix C of The Roads Board Report 'Well Maintained Highways – Code of Practice for Highway Maintenance Management'*, 1st edn. London: UK Roads Board.

2.6.5   The most recent judgement of note was *Gorringe* v. *Calderdale MBC* (2004), where a case was brought against a highway authority for failing to maintain a 'SLOW' marking on the approach to a sharp crest. The judgement confirmed a number of important points:

- the authority's duty to 'maintain' covers the fabric of a highway, but not signs and markings;
- there is no requirement for the highway authority to 'give warning of obvious dangers'; and
- drivers are 'first and foremost responsible for their own safety'.

2.6.6   Some claims for negligence and/or failure to carry out a statutory duty have been made under section 39 of the Road Traffic Act 1988,[16] which places a general duty on highway authorities to promote road safety. In connection with new roads, section 39 (3)(c) states that highway authorities 'in constructing new roads, must take such measures as appear to the authority to be appropriate to reduce the possibilities of such accidents when the roads come into use'.

2.6.7   The *Gorringe v. Calderdale* judgment made it clear that section 39 of the Road Traffic Act 1988 cannot be enforced by an individual, however, and does not form the basis for a liability claim.

2.6.8   Most claims against an authority are for maintenance defects, claims for design faults being relatively rare.

2.6.9   Advice to highway authorities on managing their risks associated with new designs is given in Chapter 5 of *Highway Risk and Liability Claims*. In summary, this advises that authorities should put procedures in place that allow rational decisions to be made with the minimum of bureaucracy, and that create an audit trail that could subsequently be used as evidence in court.

2.6.10   Suggested procedures (which accord with those set out in Chapter 3 of MfS) include the following key steps:

- set clear and concise scheme objectives;
- work up the design against these objectives; and
- review the design against these objectives through a quality audit.

2.7   Disability discrimination

2.7.1   Highway and planning authorities must comply with the Disability Equality Duty under the Disability Discrimination Act 2005.[16] This means that in their decisions and actions, authorities are required to have due regard to the six principles of:

- promote equality of opportunity between disabled persons and other persons;
- eliminate discrimination that is unlawful under the 2005 Act;
- eliminate harassment of disbled persons that is related to their disabilities;
- promote positive attitudes towards disabled persons;
- encourage participation by disabled persons in public life; and
- take steps to take account of disabled persons' disabilities, even where that involves treating disabled persons more favourably than other persons.

2.7.2   Those who fail to observe these requirements will be at the risk of a claim. Not only is there an expectation of positive action, but the duty is retrospective and local authorities will be expected to take reasonable action to rectify occurrences of non-compliance in existing areas.

2.7.3   The Disability Rights Commission (DRC) have published a Statutory Code of Practice on the Disability Equality Duty and they have also published specific guidance for those dealing with planning, buildings and the street environment.[17]

15 Highways Act 1980. London: HMSO. 16 Road Traffic Act 1988. London: TSO.
16 Disability Discrimination Act 2005. London: TSO.
17 Disability Rights Commission (DRC) (2006) *Planning, Buildings, Streets and Disability Equality. A Guide to the Disability Equality Duty and Disability Discrimination Act 2005 for Local Authority Departments Responsible for Planning, Design and Management of the Built Environment and Streets.* London: DRC.

3

The design process – from policy to implementation

Llewelyn Davies Yeang

Chapter aims

- Set out the design process in broad terms and reinforce the importance of collaborative working.

- Demonstrate the advantages of local authorities following a Development Team approach and emphasise the benefits of the developer engaging with the team at an early stage in the design process.

- Look at the key principles within the design process, and the use of design codes.

- Introduce a user hierarchy where pedestrians are considered first in the design process.

- Recommend a new approach to street and safety audits.

3.1   Introduction

3.1.1   The life of a scheme, from conception to implementation and beyond, can be broken down into seven key stages, as shown in Fig. 3.1.

3.1.2   This seven-stage process is generally applicable to all schemes, from large new developments, through to smaller infill schemes and improvements to existing streets. The key aspects are that:
-   design decisions reflect current policies;
-   policies are interpreted on a case-by-case basis and are used to define objectives; and
-   scheme designs are tested against these objectives before approval is given to their implementation.

3.1.3   The process is a general one and should be applied in a way appropriate to the size and importance of the proposal. For example, the design stage refers to the desirability of preparing a masterplan for large schemes. This is unlikely to be the case for smaller developments and improvement schemes for existing streets which are likely to be less complex, and, in some cases, a scheme layout is generally all that is required.

Figure 3.1 **The seven key stages in the life of a scheme.**

3.2   Integrated street design – a streamlined approach

3.2.1   The developer's design team needs to engage with several departments within the local planning and highway authorities in order to identify all the relevant issues. It is therefore recommended that planning and highway authorities, together with other public agencies, such as those responsible for waste collection and drainage, coordinate their activities to ensure that they do not give contradictory advice or impose conflicting conditions on the developer and the design team (Fig. 3.2).

Llewelyn Davies Yeang

Figure 3.2 **Multi-disciplinary collaborative planning helps identify all the relevant issues.**

Walsall: the Development Team approach

 **Walsall** Council

Walsall Council has successfully run a Development Team for some years. Developers submitting major planning applications benefit from meetings with officials representing a broad range of disciplines. They cover Highways, Pollution Control, Housing Services, Building Control, Development Control, Ecology, Landscape and Arboriculture (officials for these disciplines are always present), and Leisure Services, Education and the Environment Agency (officials for these disciplines are brought in as required).

From a list of available time slots at least 10 days in advance, applicants book a meeting with the Development Team, submitting their preliminary proposals at the same time. This gives ample opportunity for initial consideration of the application, including site visits if necessary.

At the meeting, developers present their proposal to the Development Team where they receive initial comments and advice. The Team provides a formal, written, fully considered response within three weeks.

Significant advantages of this approach are that the developers can plan their presentation to suit their development programme and the Team can offer advice on key elements of the proposal at an early stage, thus minimising the need for costly changes later on.

3.2.2   Local authorities should enable developers to engage effectively with individual departments by establishing a single point of contact. Some local authorities have created development teams so that all council departments with an interest in street design work together during the design and approval process (see 'Walsall case study box'). Authorities that have adopted a similar approach for larger schemes include North Somerset District Council and Oxfordshire County Council in association with the District Councils. This has clear advantages when dealing with large or small development proposals. The same approach can be adopted by local authorities internally when considering improvements to existing streets.

3.2.3   The benefits of an integrated approach applies to all stages in the process, up to and including planning how the street will be maintained in future.

3.3   Steps in the design process

3.3.1   The seven-stage process will need to be tailored to particular situations, depending on the type and complexity of the scheme. It is therefore recommended that, at the outset, a project plan is drawn up by the developer and agreed with stakeholders. The plan should include a flow chart diagram and an indication of the level and scope of information required at each stage.

3.3.2   Consultation with the public (including organisations representing particular groups) is not shown as a single, discrete stage. Public consultation should take place at appropriate points in the process. The timing and number of public consultation events will vary depending on the size and complexity of the scheme.

3.3.3   Where schemes are significant because of their size, the site or other reasons, local planning authorities and developers are encouraged to submit them to the Commission for Architecture and the Built Environment (CABE) for Design Review at the earliest opportunity.[1] Design Review is a free advice service offering expert, independent assessments of schemes.

3.3.4   Table 3.1 shows how the process can be applied. It should be noted that these steps are indicative and will vary in detail from scheme to scheme.

3.4   Stage 1: policy review

3.4.1   Street designs should generally be consistent with national, regional and local policy. The process begins with a review of relevant planning and transportation policies, and the identification of the required key design principles.

3.4.2   The starting point for the review of local policy is the Local Development Framework. The Local Transport Plan will need to be considered and authorities may also have prepared a Public

1   Communities and Local Government (2006) *Circular 1/06 Guidance on Changes to the Development Control System*. London: TSO. paragraph 76.

Table 3.1 Indicative steps in the design process for new developments and changes to existing streets

Key stages	Key activity/outputs	Responsibility	Large development	Small development	Changes to existing streets
1. Policy review	Review national, regional and local policy context	Design team	✓	✓	✓
	Review Local Transport Plan	Design team	✓		✓
	Review Public Realm Strategy	Design team	✓		✓
	Review any Street Design Guidance not included in the Local Development Framework	Design team	✓	✓	✓
2. Objective setting	Prepare Development Brief	Planning and highway authorities	✓		
	Agree objectives	All	✓	✓	✓
3. Design	Carry out context appraisal	Design team	✓	✓	✓
	Develop proposed movement framework	Design team	✓	✓	
	Prepare outline masterplan or scheme layout	Design team, working closely with other stakeholders	✓	✓	✓
	Develop street character types	Design team	✓	✓	✓
	Design street network	Design team	✓	✓	
	Produce detailed masterplan or scheme layout	Design team	✓	✓	✓
	Produce design code	Design team	✓		
4. Quality auditing	Carry out particular audits required to assess compliance with objectives	Prepared by design team, considered by planning and highway authorities	✓	✓	✓
5. Planning approval	Prepare design and access statement and other supporting documents	Prepared by design team for approval by the planning authority in consultation with the highway authority	✓	✓	
	Outline planning application		✓	✓	
	Full planning application		✓	✓	
6. Implementation	Detailed design and technical approval	Design team	✓	✓	✓
	Construction	Promoter	✓	✓	✓
	Adoption	Highway authority	✓	✓	
7. Monitoring	Travel plan	Promoter	✓		
	Road user monitoring	Highway authority			✓

Realm Strategy or Open Space Strategy which will be of particular importance in establishing fundamental design principles. The policy review should also consider the national policy framework, particularly where the local policy framework is out of date or unclear.

3.5   Stage 2: objective setting

3.5.1   It is important that objectives for each particular scheme are agreed by all parties and reviewed later in the process to ensure that they are being met. Objectives need to reflect the local policies and the wider planning framework to ensure a consistency of approach across an area.

3.5.2   On complex and lengthy projects, objectives may need to be reviewed and revised as the design process proceeds, with any changes agreed by all parties.

3.5.3   Objectives should be expressed as outcomes that can be readily measured, and should not be expressed in vague terms, or require or invoke particular solutions. The objectives will often be related to the various activities expected to take place in particular locations and streets. There may also be objectives that apply across the whole of a new development area.

3.5.4   Typical objectives might be:
- enabling local children to walk and cycle unaccompanied from all parts of a development to a school, local park or open space;
- promoting and enhancing the vitality and viability of a local retail centre;

- ensuring that a development will be served by public transport that is viable in the long term; and
- keeping traffic speeds at 20 mph or less in all streets on a development.

3.5.5   Objectives could be expressed as a design checklist, which provides a simple summary of the key aspects that need to be met.

3.5.6   For some sites, a Development Brief or other form of guidance may have been prepared to establish the key principles of development, and will need to be taken into account at the objective setting stage.

3.6   Stage 3: design

Context appraisal

3.6.1   A context appraisal will normally be undertaken to determine how buildings and streets are arranged within the local area. This will be used to help determine an appropriate form for the development of, or changes to, existing streets.

3.6.2   The context appraisal will identify how an area has developed in terms of form, scale, the pattern and character of streets and how a site or existing street relates to existing buildings and/or open space. It may also be appropriate to identify poor-quality streets or areas which need to be improved. One way of achieving this is by undertaking a Landscape Character Appraisal.[2]

3.6.3   On smaller schemes it may only be necessary to consider context in a relatively local area, but this does not prevent designers from drawing on good-quality examples of local distinctiveness from the wider area.

2 Countryside Agency and Scottish Natural Heritage (2002) *Landscape Character Assessment: Guidance for England and Scotland*. London: TSO.

Figure 3.3 New housing with: (a) good (b) poor integration into an existing street.

Development opportunity sites	
View towards the river	– →
New active frontage onto London Road	↓↓
Conservation area	
Character buildings	▱▱▱
Green network	– →
Major riverside green link/space (pedestrian)	▪
New aspect onto river	→
Pedestrian links from station/interchange	– →
New street with possible bridge over railway	⇔→
Residential (existing)	
Employment & 'consultation zone'	⧄
Existing vegetation	....
Mixed use, higher density, centre focus	◌
Railway station / interchange	⫽

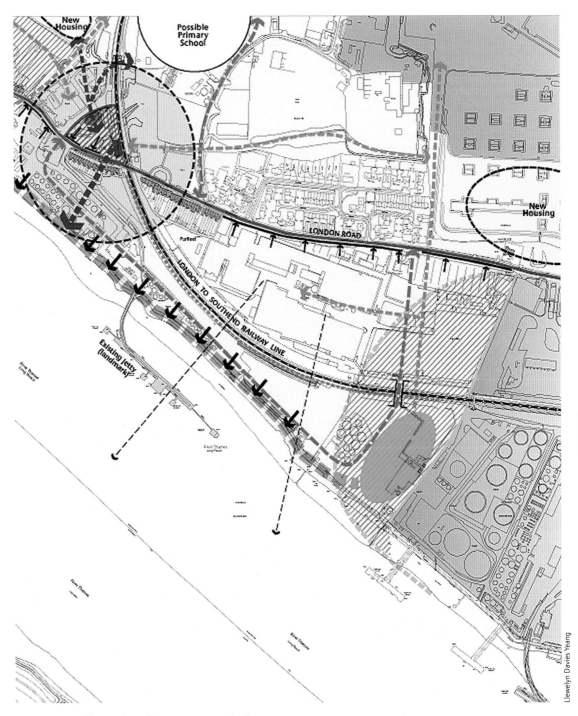

Figure 3.4 An illustration of a context appraisal.

3.6.4   When existing streets are being redesigned, it is very important to have a detailed understanding of how they sit within an urban area. Care needs to be taken to retain and develop the relationship between the streets and the buildings and public spaces that surround them, and to capitalise on links to important local destinations. There is a need to identify opportunities to repair incomplete or poor-quality connections (Fig. 3.3).

Analysis of existing places

3.6.5   As part of the context appraisal, the relative importance of existing places within the locality will need to be identified. Places to be identified include important buildings and public open spaces, and key destinations such as educational institutions and areas of employment or commerce (Fig. 3.4).

3.6.6   The analysis will determine which places in the surrounding area need to be made accessible to local people, particularly on foot and by bicycle, and the appropriate design and layout of that area.

3.6.7   This analysis will also help to establish whether additional centres of activity are required as part of a new development, such as a new local centre or school.

Analysis of existing movement patterns

3.6.8   It is recommended that the design of a scheme should follow the user hierarchy shown in Table 3.2.

Table 3.2: User hierarchy

Consider first	Pedestrians
	Cyclists
	Public transport users
	Specialist service vehicles (e.g. emergency services, waste, etc.)
Consider last	Other motor traffic

3.6.9   The hierarchy is not meant to be rigidly applied and does not necessarily mean that it is always more important to provide for pedestrians than it is for the other modes. However, they should at least be considered first, followed by consideration for the others in the order given. This helps ensure that the street will serve all of its users in a balanced way. There may be situations where some upper-tier modes are not provided for – for example, buses might not need to be accommodated in a short, narrow section of street where access for cars is required.

3.6.10   An analysis of movement within an existing settlement will help identify any changes required for it to mesh with a new development. It could also influence movement patterns required within the new development.

3.6.11   The position of a street within the existing movement framework will determine the demands it needs to meet, and these, in turn, will inform decisions on its capacity, cross-section and connectivity.

3.6.12   Establishing the movement requirements of existing streets is particularly important when changes are planned so that the needs of all road users are fully taken into account.

Proposed movement framework

3.6.13   For new developments, an understanding of how an existing area functions in terms of movement and place enables the proposed points of connection and linkage to be identified, both within and from the site, so that important desire lines are achieved. This process will help ensure that a new development enhances the existing movement framework of an area rather than disrupting or severing it (Fig. 3.5).

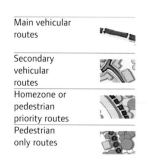

Main vehicular routes

Secondary vehicular routes

Homezone or pedestrian priority routes

Pedestrian only routes

Llewelyn Davies Yeang

Figure 3.5 Proposed movement diagram for the redevelopment of RAF Halton.

Figure 3.6 A concept masterplan with 3-D visualisation.

3.6.14   Guidance on the design of movement frameworks is set out in more detail in Chapter 4. The movement framework is a key input to the development of the masterplans.

Outline masterplan or scheme layout

3.6.15   Although not always needed, especially where proposals are small scale, an *outline masterplan* helps to establish the scheme's broad development principles (Fig. 3.6).

3.6.16   An outline masterplan that has been produced through collaboration with key stakeholders is usually more robust and realistic than it would otherwise be. For larger sites, a series of stakeholder events is often the most productive way of achieving this as it brings all the parties together to generate a design vision which reflects community and stakeholder objections. For smaller sites, the process need not be so involved and design proposals may be more appropriately informed by a simple scheme layout developed though targeted meetings with key stakeholders and/or correspondence.

3.6.17   For simpler schemes adequately served by detailed layouts, outline scheme layouts are usually not likely to be needed (Fig. 3.7). An exception might be where, for example, the site is in a conservation area.

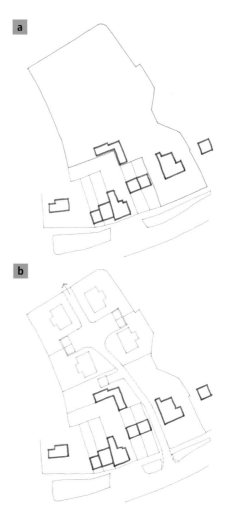

Figure 3.7 Small scheme design for an infill development (a) outlined in red. Location of new houses (b) shown in green together with new access street. Note that the new access street can be extended to allow for future growth at the top of the diagram.

3.6.18   The outline masterplan will bring together the movement framework with other important aspects of the design of a new development, such as the need for new local facilities, important views and microclimate considerations.

3.6.19   When developing outline masterplans for large-scale proposals, such as an urban extension, the design team needs to consider the longer-term vision for the area in question. Such a future-proofing exercise involves looking beyond the usual planning periods to consider where development may be in, say, 20 or 30 years. The issues identified may influence the masterplan. An example would be allowing for the future growth of a settlement by continuing streets to the edge of the site so that they can be extended at a later date (Fig. 3.8). This principle also applies to smaller-scale schemes which need to take account of future development proposals around an application site and, where appropriate in discussions with the local planning authority, to ensure that linkages are established wherever possible and that the site is swiftly integrated into its surroundings.

Street character types

3.6.20   Once the outline masterplan has been prepared, the next step will be to establish the characteristics of the various types of street that are required for the new development.

3.6.21   Street character types set out not only the basic parameters of streets, such as carriageway and footway widths, but also the street's relationship to buildings and the private realm, and other important details, such as parking arrangements, street trees, planting and lighting.

3.6.22   Further guidance on determining street character types is given in Chapter 7.

3.6.23   Street character types can also be expressed through design codes, which are discussed later in this chapter.

Street network

3.6.24   It is recommended that the proposed street network is based on a combination of the proposed movement framework and the proposed street types (Fig. 3.9).

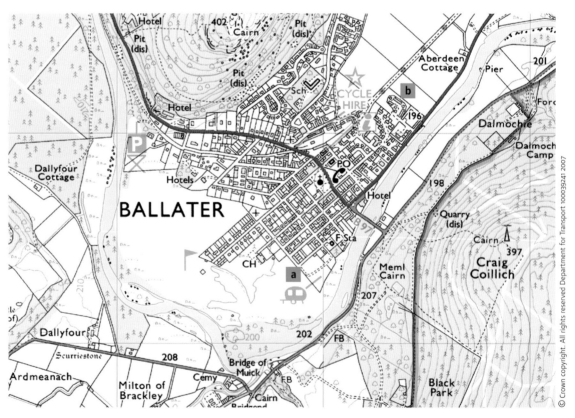

Figure 3.8 Ballater, Aberdeenshire – the ability for future growth is not compromised in the south-west of the village (a) with its permeable street pattern, but more recent cul-se-sac type development in the north-east (b) does not allow for a connected growth of the village.

Detailed masterplan or detailed scheme layout

3.6.25   *Detailed masterplans* are likely to be needed for schemes at the higher end of the scale in terms of size and complexity. For relatively simple proposals, a detailed scheme layout is all that is likely to be needed. Guidance on the masterplanning process is given in *Creating Successful Masterplans: A Guide for Clients*.[3]

3.6.26   It is important when preparing a detailed masterplan, that all of the critical features which impact on the efficiency and quality of the development – and which cannot be changed once it is built – are carefully considered (Fig. 3.10).

3.6.27   The full extent of the masterplanning process is beyond the scope of MfS, but it is recommended that the following key features relating to street design are addressed:
* connections to the surrounding area;
* connections through the site;
* street layout and dimensions;
* building lines;
* building heights;
* routes for utilities;

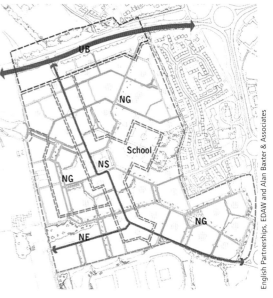

Figure 3.9 Street network diagram for Upton, Northamptonshire, showing the main route through a connected layout and linkages to key spaces and places within the development, with street character types identified.

* parking provision, design and control;
* landscape design and structural planting;
* materials, management and maintenance regime;
* servicing and access for emergency vehicles;
* speed control; and
* SUDS and sewer routes.

Figure 3.10 An example of a large-scale masterplan – Sherford New Community near Plymouth.

3   CABE (2004) *Creating Successful Masterplans: A Guide for Clients*. London: CABE

Design codes

3.6.28   Design codes are an effective mechanism for implementing the masterplan (Fig. 3.11). They comprise detailed written and graphically presented rules for building out a site or an area. They are often promoted by local authorities but they may be put forward by the private sector.

3.6.29   Design codes determine the two- and three-dimensional design elements which are key to the quality of a development. Although not appropriate in all circumstances, they can be valuable for helping local authorities and developers to deliver high-quality design.

3.6.30   The elements which are coded will differ according to circumstances, but they might include aspects relating to layout, townscape

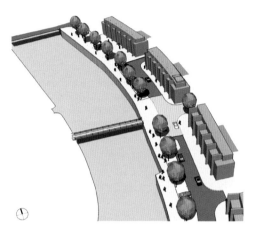

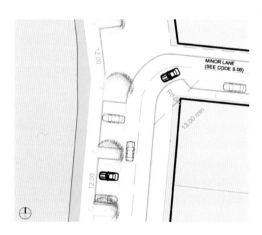

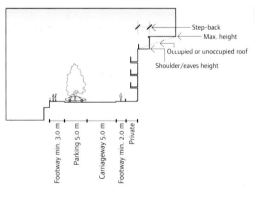

S.04.1 River's edge · Plot series	
Plot series	**Riverside** **Character area**
Attribute	**Code**
Priority relative to street	Active frontage must be orientated toward the street
Sub-series	
Series type (regular/mixed)	Regular
Plot width at frontage (dimension range)	6 m – 12 m; exceptionally, larger width, in increments of 5 m, with vertical articulation of module visible in the façade
Point of access (type and frequency)	
Pedestrian	Minimum every 12 m
Vehiclar	None
Allowable plot types	Attached

S.04.2 River's edge · Plan	
Public highway	**Riverside** **Character area**
Attribute	**Code**
Carriage width	6.0 m
Footpath width	Min 2.0 m, Min 3.0 m along riverfront
Design speed	20 mph
Traffic calming	Carriageway narrowing
Junction radii	Min. 40 m
Vehicle type to be accommodated	Cars, small service vehicles, fire appliances, cycles
On-street parking	Perpendicular – 5.0 m x 2.5 m
Direct access to plots	No
Street trees	8.0 – 12.0 m spacing between trees (adjust to accommodate parking areas)

S.04.3 River's edge · Section	
Street section	**Riverside** **Character area**
Attribute	**Code**
Shoulder/eaves height	3 – 5 storey
Storey height	Floor to ceiling heights on the ground floor must be a minimum of 2.7m to allow for flexibility of use and adaptability
Maximum height of roof occupied or unoccupied	3 m above shoulder/eaves height
Step-back	2.5 m maximum
Balconies	1.5 m maximum
Vertical position of access	Level
Vertical mix of uses	Residential on ground floor, retail on ground floor around junctions of Burrell Street and Water Lane, residential above ground floor

Figure 3.11 Design code for riverside development in Rotherham.

Roger Evans & Associates

and landscape considerations, or architecture or building performance. Codes may also usefully establish the relationships of plots, sometimes the building form or even materials. However, given the primary need to secure a quality townscape and a sense of place, the most important role of a design code will be in securing the lasting structural elements of a place, such as the street pattern and street dimensions. Getting these structural elements right will enable the other elements of a design to evolve. To do this successfully, however, the design code will need to be underpinned by a specific design vision, such as a masterplan or a design and development framework.

3.6.31   A key benefit of design codes is the collaborative nature of their preparation – a process that brings together a broad range of professionals and organisations with a role in delivering the development. Typically, this comprises land, design, development and public interests. Regardless of whether a code is promoted by the private sector or a local authority, it is essential that engineers, designers and planners work together to develop the code to help ensure that each aspect of the design successfully reinforces the overall sense of place.

3.6.32   When a code is prepared by a local authority, a Development Team approach will bring advantages. Representatives from the authority's key departments will need to work together. These will include planning (both policy and development control), highways, landscape, parks and recreation, and, where appropriate, the housing authority and the authority's estates

management team. The inclusion of the authority's legal team will also be helpful, particularly where the codes relate to planning conditions, section 106 and 278 agreements, unilateral undertakings or local development orders. In particular, the highways team in an authority plays a key role in the preparation of a design code and in adopting the infrastructure that results.

3.6.33   Detailed guidance on the preparation and implementation of design codes, including advice on how they can be formalised, is set out in *Preparing Design Codes – A Practice Manual*.[4] This guidance makes it clear that:

> 'Highways policy and standards are decisive influences on design code preparation, and design codes provide a key opportunity to improve highways design that takes account of urban design considerations and helps create quality places. The preparation of a design code can provide a ready opportunity to work closely with highways authorities to review any outdated local highways standards.'

3.6.34   In this context it is essential that, when design codes are being prepared, the coding team consider carefully what the design objectives are and the required outcomes to deliver those objectives. It is recommended that careful consideration should be given to the scope for the design code to address those aspects of the street environment that will be crucial to delivering the required outcomes. Those which are not can be left to the discretion of the developer and his or her designer (see box and Fig. 3.12).

Design codes

Street-related design elements and issues which a design code may relate to include:
- the function of the street and its position in the Place and Movement hierarchy, such as boulevards, high streets, courtyards, mews, covered streets, arcades or colonnades;
- the principal dimensions of streets;
- junctions and types of traffic calming;
- treatments of major junctions, bridges and public transport links;

- location and standards for on-and off-street parking, including car parks and parking courts, and related specifications;
- street lighting and street furniture specifications and locations;
- specifications for trees and planting;
- location of public art;
- drainage and rainwater run-off systems;
- routeing and details of public utilities; and
- arrangements for maintenance and servicing.

4   Communities and Local Government (2006) *Preparing Design Codes – A Practice Manual.* London: RIBA Publishing.

a Criteria	Street Specification		
	Standard Design	Variation 1 (One-sided parking)	Variation 2 (Variable Kerb)
Design Speeds			
Speed Limit	20 mph (at entrance)		
Control Speed	20 mph (internally)		
Street dimensions and character			
Minimum carriageway width	5.5 m		
Footway	2.0-3.0 m on each side		
Cycle way	No - Parallel routes provided on other streets		
Verge	No		
Private strip	2.0 m		
Direct vehicular access to properties	Yes		
Plot Boundary Treatment	2.0 m private area to building line with up to 1.0 m encroachment 0.9-1.1 m railing on plot boundary with footway		
Maximum number of properties served	Not restricted		
Public Transport			
Bus access	No		
Street design details			
Pull out strip	No		
Traffic calming	Features at 60 m-80 m c/c, parking, trees, formal crossings		Non parallel kerbs, variations in planting/ building lines, parking
Vehicle swept path to be accommodated	Removals/refuse vehicles enter and leave using own side of road only (assuming 20 mph)		Refuse vehichle passing car on street
On street parking	Yes, both sides, 2.0 m wide	Yes, one side, 2.0 m wide	Yes, one or both sides, informal
Gradients (footways)	1:15 Maximum, footway to follow carriageway		
Maximum foward visibility	33 m, 20 m (measured 1.0m out from kerb)		
Junction sightlines (x/y)	2.4 m/33 m		
Junction spacing-same side/other side	60 m/30 m		
Junction radii	4 m		
Stats services (excluding storm and capping layer drainage	In footway, each side. Drainage below carriageway	Footways, where necessary	
Materials			
Footway Surfacing	· Natural grey, pre-cast concrete paving flags, 63 mm thick staggered joint, variable sizes: 600x450 mm, 450x450 mm-10%, 300 x 450 mm		
Parking Zone	· Natural grey tumbled pre-cast concrete paviors 80 mm thick with 225-300 mm exposed granite aggregate pre-cast kerb 20 mm high	n/a	
Kerbing	· 225-300 mm wide x 200 mm square edged exposed granite aggregate pre-cast kerb 125 mm high · 225-300 mm wide x 200 mm square edged exposed granite aggregate pre-cast kerb 20 mm high		
Carriageway	· Black-top		
	· 5 rows of 100 mm x 100-250 mm cropped granite setts		
Pedestrian Crossing	· Stainless steel tactile studs inserted into paving/tactile paving	Tactile Paving	
Street Lighting	LC4, LC5 Maximum to eaves height (see Appendix 4)		
Street Furniture	SF3, SF6, SF9 (see Appendix 4)		
Trees			
Street Trees	Acer platanoides 'Obelisk'		
Feature Trees	Corylus Colurna - specific locations detailed in Development Briefs		

English Partnerships, EDAW and Alan Baxter Asssoicates

Figure 3.12 (a) and (b) Design code for particular street character type in Upton, Northampton (note (b) is on the next page).

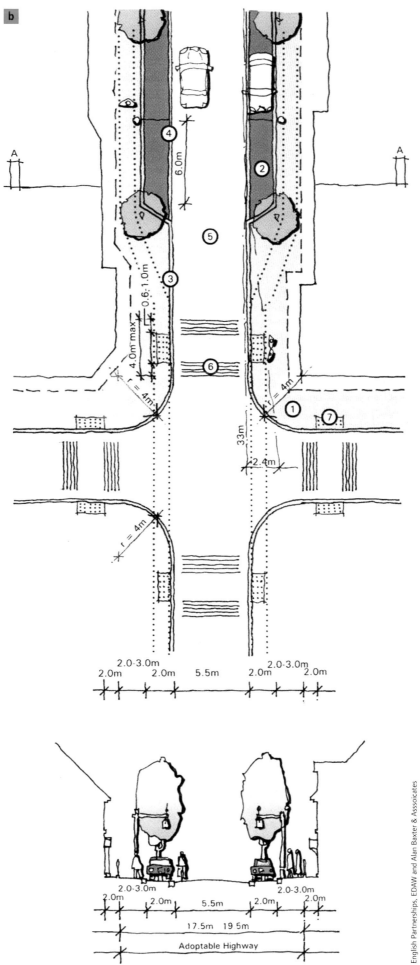

b

2.0-3.0m
2.0m 2.0m 5.5m 2.0m 2.0-3.0m
                              2.0m

2.0-3.0m
2.0m 2.0m 5.5m 2.0m 2.0-3.0m
                              2.0m

17.5m   19.5m

Adoptable Highway

English Partnerships, EDAW and Alan Baxter & Asssoicates

3.7 Stage 4: quality auditing

3.7.1 Properly documented design audit and sign-off systems are important. They help ensure that street designs are appropriate and meet objectives agreed at the outset. Such audits may include documents required by the local planning authority to support an outline or detailed application. In existing streets, quality audits provide an opportunity for decision makers to make a balanced assessment of different considerations before approving a particular solution (see 'Devon case study box').

3.7.2 Being made up of a series of assessments, a quality audit is likely to be carried out by various professionals and each may be undertaken within particular guidelines. By grouping the assessments together, any compromises in the design will be apparent, making it easier for decision makers to view the scheme in the round.

3.7.3 Auditing should not be a box ticking exercise. It is an integral part of the design and implementation process. Audits inform this process and demonstrate that appropriate consideration has been given to all of the relevant aspects. The quality audit may include some or all of the following, or variations on them, depending on the nature of the scheme and the objectives it is seeking to meet:

- an audit of visual quality;
- a review of how the streets will be used by the community;
- a road safety audit, including a risk assessment (see below);
- an access audit;
- a walking audit;[5]
- a cycle audit;[6, 7]
- a non-motorised user audit;[8]
- a community street audit (in existing streets);[9] and
- a Placecheck audit.[10]

3.7.4 Access auditors should take account of the advice given in *Inclusive Mobility*.[11] The Centre for Accessible Environments has also published guidance on access audits in relation to public buildings.[12] It contains much useful general advice on access auditing in the public realm.

5 PERS (Pedestrian Environment Review System) is software developed by TRL and provides one way of carrying out a walking audit. For further details see www.tfl.gov.uk/ streets/walking/reports.shtml.
6 TRL (unpublished) *Cycle Environment Review System*.
7 Institution of Highways and Transportation (IHT) (1998) *Cycle Audit and Cycle Review*. London: IHT.
8 Highways Agency (HA) (2005) HD42 *Non-Motorised User Audits – Volume 5 Sections 2 Part 5. Design Manual for Roads and Bridges*. London: TSO.
9 Living Streets (2003) *DIY Community Street Audit Pack*. London: Living Streets.
10 Guidance on Placecheck is available at www.placecheck.info.
11 Department for Transport (2002) *Inclusive Mobility A Guide to Best Practice on Access to Pedestrian and Transport Infrastructure*. London: Department for Transport.
12 Centre for Accessible Environments (2004) *Designing for Accessibility*. London: RIBA Publishing.
13 IHT (1996) *The Safety Audit of Highways*. London: IHT.
14 HA (2003) HD19 *Road Safety Audit – Volume 5 Section 2 Part 2. Design Manual for Roads and Bridges*. London: TSO.

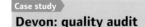
Devon: quality audit

Figure 3.13 Road safety officers, police and engineers working on a road safety audit in Devon.

Devon County Council has developed a process whereby both an environmental audit and a road safety audit (Fig. 3.13) are carried out when improvement schemes are being prepared.

The two audits are carried out separately and if there is a difference of opinion between the two over any aspect, the matter is referred to a senior officer for a decision. It is therefore possible to demonstrate that decisions have been properly considered in case of future dispute.

This process is, in essence, a quality auditing process.

Road safety audits

3.7.5 Road safety audits (RSAs) are routinely carried out on highway schemes. The Institution of Highways and Transportation (IHT) Guidelines on RSA[13] sit alongside the Highways Agency standard contained in DMRB[14] as the recognised industry standard documents in the UK. The procedures set out in DMRB are a formal requirement only for trunk roads.

3.7.6 RSAs are not mandatory for local highway authorities. Many residential streets, where the design is carried out by a developer's consultant, are assessed independently by the local highway authority. In some authorities there is no requirement for a further check by an RSA team, particularly where it is clear that motorised traffic volumes and speeds, and the degree of potential conflict between different user-groups, are not going to be significant.

Manual for Streets

3.7.7   The purpose of the RSA is to identify road safety problems, with the objective of minimising the number and severity of casualties. An RSA is not a check on compliance with design standards. Audits take all road users into account, including pedestrians and cyclists. The standard procedure is that the auditor makes recommendations for changes to the design to address perceived safety concerns. The design team reviews the RSA report and decides whether or not to accept particular recommendations.

3.7.8   *It is important to note that the design team retains responsibility for the scheme, and is not governed by the findings of the RSA.* There is therefore no sense in which a scheme 'passes' or 'fails' the RSA process. Designers do not have to comply with the recommendations of a safety audit, although in such cases they would be expected to justify their reasoning in a written report.

3.7.9   The process set out in DMRB requires the audit team to be independent of the design team. Road safety issues are therefore often considered in isolation from visual quality and Placemaking issues, and it can be difficult to achieve a balanced design through dialogue and compromise. However, the requirement for independence need not prevent contact between the design team and the audit team throughout the process.

3.7.10   It is beyond the scope of MfS to define in detail a wholly new and more balanced approach to RSAs, and the IHT guidelines are due to be revised. However, involving road safety professionals as an integral part of the design team could help to overcome some of the reported problems. This allows ideas to be tested and considered in more balanced and creative ways.

3.7.11   One area of concern with the existing system is that RSAs may seek to identify all possible risks without distinguishing between major and minor ones, or quantifying the probability of them taking place. There can also be a tendency for auditors to encourage designs that achieve safety by segregating vulnerable road users from road traffic. Such designs can perform poorly in terms of streetscape quality, pedestrian amenity and security and, in some circumstances, can actually reduce safety levels.

3.7.12   It would therefore be useful if RSAs included an assessment of the relative significance of any potential safety problems. A risk assessment to consider the severity of a safety problem and the likelihood of occurrence would make it considerably easier for decision makers to strike an appropriate balance. An example of a risk assessment framework is given in *Highway Risk and Liability Claims*.[15]

3.1.13   Careful monitoring (such as through conflict studies) of the ways in which people use the completed scheme can identify any potential safety problems. This can be particularly useful when designers move away from conventional standards. Monitoring is discussed further in Section 3.10 below.

3.8   Stage 5: planning approval

3.8.1   New development proposals need to be submitted for approval to the planning authority who, in turn, consults with the local highway authority on street design issues.

3.8.2   Where outline planning permission is being sought, various supporting information needs to be provided as agreed with the planning and highway authorities. This may include some or all of the following, depending on the type size and complexity of the scheme (this list is not necessarily exhaustive):
·   preliminary street designs and layouts;
·   a Design and Access Statement (see box);[16, 17, 18]
·   a Transport Assessment;
·   a Travel Plan;
·   an Environmental Statement or Environmental Impact Assessment;
·   a Sustainability Appraisal;
·   a Flood Risk Assessment; and
·   a Drainage Report.

Design and Access Statement

Since August 2006, Design and Access Statements (DASs) have been required for most planning applications for new developments.[19] DASs are documents that explain the design thinking behind a planning application and are therefore important documents. They normally include a written description and justification of the planning application, often using photographs, maps and drawings to help clarify various issues.

15 UK Roads Board (2005) *Highway Risk and Liability Claims – A Practical Guide to Appendix C of The Roads Board Report 'Well Maintained Highways – Code of Practice for Highway Maintenance Management'*, 1st edn. London: UK Roads Board.
16 Communities and Local Government (2006) *Circular 01/06 Guidance on Changes to the Development Control System*. London: TSO.
17 CABE (2006) *Design and Access Statements – How to Write, Read and Use Them*. London: CABE.
18 Disability Rights Commission (DRC) (2005) *Planning, Buildings, Streets and Disability Equality*. Stratford upon Avon: DRC.
19 ibid. (16).

3.8.3   It is critical that as many issues as possible are resolved at the outline planning application stage so that they can receive thorough and timely consideration. This will help to make detailed planning applications or the consideration of reserved matters as straightforward as possible.

3.8.4   The local planning authority needs to ensure that the key features set out in paragraph 3.6.27 above, and any site-specific issues of importance, are resolved before outline permission is granted. The design of streets, spaces and parking is important and should be considered alongside other planning matters, such as the design of the built form and use, conservation, landscape and housing type.

3.8.5   Ideally, following outline consent, only matters of detail, such as detailed layout and material choices, will be left for consideration at the detailed application stage.

3.8.6   For small developments and schemes in sensitive locations, such as conservation areas, it will often be appropriate for detailed planning approval to be sought without first obtaining outline consent. This enables the approving authorities to consider the effects of the development in detail before approving the development in principle.

3.8.7   In existing streets, the highway authority is normally both the designer and the approving body. It is recommended that well-documented approval systems are used that properly assess the impact of proposed changes to prevent the gradual degradation of the street scene through ill-considered small-scale schemes.

3.9   Stage 6: implementation

Detailed design, technical approval, construction and adoption

3.9.1   In the past, developers have sought to satisfy the detailed planning process before commencing the detailed design of streets in order to meet the highway adoption process. This has led to problems in some circumstances where the detailed design and technical approval process throws up problems that can only be resolved by changing the scheme that was approved at the detailed planning stage.

3.9.2   A more integrated approach is recommended, with highway adoption engineers being fully involved throughout, so that schemes that are approved at detailed planning stage can move through the technical approval stage without requiring any significant changes. Highway adoption is dealt with in more detail in Chapter 11.

3.10   Stage 7: monitoring

3.10.1   Planning Policy Statement 3: Housing (PPS3)[20] makes clear that local planning authorities and agencies are expected to report on progress towards the achievement of consistently good design standards through the Annual Monitoring Report process, assessing achievement against their design quality objectives (PPS3, paragraphs 75–77). This is likely to include some consideration of the design quality of new streets or existing street modifications as part of the wider public realm.

3.10.2   Monitoring is an integral element of the disability equality duty under the Disability Discrimination Act 2005.[21] Within their Disability Equality Schemes, local authorities are expected to set out their arrangements for monitoring the effectiveness of their policies and practices as they relate to the interests of disabled people. This includes both planning and highways functions. The information will help authorities to make decisions about what actions and changes to their policies and practices would best improve disability equality.

3.10.3   Monitoring for reasons other than those above has seldom been undertaken to date but can be highly desirable. Monitoring can be used to see how completed schemes or existing street environments function in practice, so that changes can be made to new designs, particularly innovative ones, at an early stage.

3.10.4   Monitoring can also be an important aspect of residential travel plans, where patterns of movement are reviewed against planned targets.

20 Communities and Local Government (2006) *Planning Policy Statement 3: Housing*. London. TSO.
21 Disability Discrimination Act 2005. London: TSO.

B

Design principles

4.1   Planning for things you cannot easily change later

4.1.1   The way streets are laid out and how they relate to the surrounding buildings and spaces has a great impact on the aesthetic and functional success of a neighbourhood. Certain elements are critical because once laid down, they cannot easily be changed. These issues are considered in the masterplanning and design coding stage, and need to be resolved before detailed design is carried out.

4.1.2   This chapter highlights the issues likely to be encountered in developing detailed designs, and ways of dealing with them. There are also tips on avoiding unwanted consequences of particular design decisions.

4.2   The movement framework

4.2.1   A key consideration for achieving sustainable development is how the design can influence how people choose to travel. Designers and engineers need to respond to a wide range of policies aimed at making car use a matter of choice rather than habit or dependence. Local transport plans and movement strategies can directly inform the design process as part of the policy implementation process (*Wales*: Regional Transport Plans and Local Development Plans).

4.2.2   It is recommended that the movement framework for a new development be based on the user hierarchy as introduced in Section 3.6. Applying the hierarchy will lead to a design that increases the attractiveness of walking, cycling and the use of public transport. Delays to cars resulting from adopting this approach are unlikely to be significant in residential areas. The movement framework should also take account of the form of the buildings, landscape and activities that form the character of the street and the links between new and existing routes and places (Fig. 4.1).

4.2.3   Street networks should, in general, be connected. Connected, or 'permeable', networks encourage walking and cycling, and make places easier to navigate through. They also lead to a more even spread of motor traffic throughout the area and so avoid the need for distributor roads with no frontage development. Research[2] shows that there is no significant difference in collision risk attributable to more permeable street layouts.

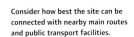

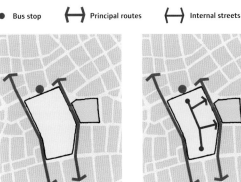  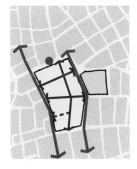 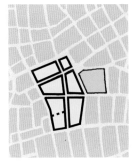

Consider how best the site can be connected with nearby main routes and public transport facilities.

The typical cul-de-sac response creates an introverted layout which fails to integrate with its surroundings.

A more pedestrian friendly approach that integrates with the surrounding community. It links existing and proposed streets and provides direct routes to bus stops.

This street pattern then forms the basis for perimeter blocks which ensure that buildings contribute positively to the public realm.

Figure 4.1 Integrating new developments into the existing urban fabric is essential (source: *The Urban Design Compendium*[1]).

1   Llewelyn Davies (2000) *The Urban Design Compendium*. London: English Partnerships and The Housing Corporation.

2   I York, A Bradbury, S Reid, T Ewings and R Paradise (2007) *The Manual for Streets: Redefining Residential Street Design*. TRL Report No. 661. Crowthorne: TRL.

4.2.4   Pedestrians and cyclists should generally be accommodated on streets rather than routes segregated from motor traffic. Being seen by drivers, residents and other users affords a greater sense of security. However, short pedestrian and cycle-only links are generally acceptable if designed well. Regardless of length, all such routes in built-up areas, away from the carriageway, should be barrier-free and overlooked by buildings. Narrow routes hemmed in by tall barriers should be avoided as they can feel claustrophobic and less secure for users.

Connecting layouts to their surroundings

4.2.5   Internal permeability is important but the area also needs to be properly connected with adjacent street networks. A development with poor links to the surrounding area creates an enclave which encourages movement to and from it by car rather than by other modes (Fig. 4.2).

4.2.6   External connectivity may often be lacking, even where layouts generally have good internal permeability. Crown Street, Glasgow, is shown in Fig. 4.3, with an indication of where connectivity was not realised as may have been intended in the masterplan.

4.2.7   The number of external connections that a development provides depends on the nature of its surroundings. Residential areas adjacent to each other should be well connected.

4.2.8   To create a permeable network, it is generally recommended that streets with one-way operation are avoided. They require additional signing and result in longer vehicular journeys.

The hierarchies of provision

4.2.9   If road safety problems for pedestrians or cyclists are identified, conditions should be reviewed to see if they can be addressed, rather than segregating these users from motorised traffic. Table 4.1 suggests an ordered approach for the review.

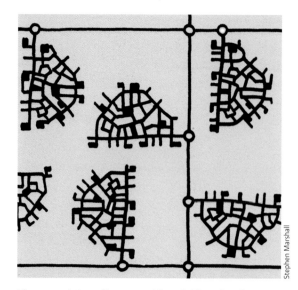

Stephen Marshall

Figure 4.2 Internally permeable neighbourhoods lacking direct connections with one another (source: Marshall 2005 [3]).

4.2.10   These hierarchies are not meant to be rigidly applied and there may be situations where it is sensible to disregard some of the solutions when deciding on the optimum one. For example, there would be no point in considering an at-grade crossing to create a pedestrian/ cyclist link between developments on either side of a motorway. However, designers should not dismiss out of hand solutions in the upper tier of the hierarchy.

4.2.11   It is recommended that the hierarchies are used not only for a proposed scheme but also for connections through existing networks to local shops, schools, bus stops, etc.

4.3   Building communities to last

4.3.1   Good design is a key element in achieving the Government's aim to create thriving, vibrant, sustainable communities. Sustainable communities meet the diverse needs of existing and future residents, are sensitive to their environment by minimising their effect on climate change, and contribute to a high quality of life. They are safe and inclusive, well planned and promote social inclusion, offering equality of opportunity and good services for all.

3   Marshall, S. (2005)
    Streets and Patterns.
    London: Spon Press.
    Figure 2.10, p.34.

Figure 4.3 Crown Street, Glasgow: (a) the Crown Street development in the background is separated from the main road to the city centre; and (b) map.

Table 4.1 The hierarchies of provision for pedestrians and cyclists

	Pedestrians	Cyclists
Consider first ↓	Traffic volume reduction	Traffic volume reduction
	Traffic speed reduction	Traffic speed reduction
	Reallocation of road space to pedestrians	Junction treatment, hazard site treatment, traffic management
	Provision of direct at-grade crossings, improved pedestrian routes on existing desire lines	Cycle tracks away from roads
Consider last	New pedestrian alignment or grade separation	Conversion of footways/footpaths to adjacent-* or shared-use routes for pedestrians and cyclists

* Adjacent-use routes are those where the cyclists are segregated from pedestrians.

4.3.2   Areas of local amenity should be more evenly distributed, with good connectivity, so that the overall layout encourages access by walking or cycling, and shortens the distances travelled by car (Fig 4.4).

4.3.3   When considering a site there needs to be a broad understanding of its historic development and its relationship with other communities, whether at the village, town or city scale (Fig 4.5).

4.3.4   The provision and viability of facilities needs to be assessed in relation to the location and scale of proposals. In many cases, it may be better for a new development to reinforce existing centres and facilities rather than providing alternative facilities. The greater the density of development, the more facilities can be supported.

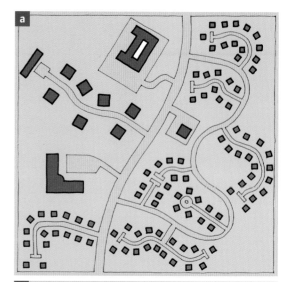

Figure 4.4 (a) dispersed and car-dependent versus (b) traditional, compact and walkable layout.

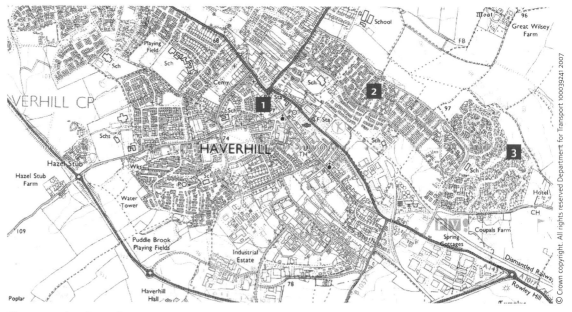

Figure 4.5 The plans of many UK villages, towns and cities illustrate different patterns of development over time, from (1) historic cores, through to (2) experimental 'Radburn' layouts from the 1960s, to (3) recent cul-de-sac/DB32-type layouts.

Manual for Streets

Figure 4.6 Perimeter blocks enclosing a pleasant communal open space.

Figure 4.7 A highways-dominated layout with buildings that have a poor relationship to the road.

4.4 The walkable neighbourhood

4.4.1   Walkable neighbourhoods are typically characterised by having a range of facilities within 10 minutes' (up to about 800 m) walking distance of residential areas which residents may access comfortably on foot. However, this is not an upper limit and PPS13[4] states that walking offers the greatest potential to replace short car trips, particularly those under 2 km. MfS encourages a reduction in the need to travel by car through the creation of mixed-use neighbourhoods with interconnected street patterns, where daily needs are within walking distance of most residents.

4.4.2   By creating linkages between new housing and local facilities and community infrastructure, the public transport network and established walking and cycling routes are fundamental to achieving more sustainable patterns of movement and to reducing people's reliance on the car. A masterplan (or scheme layout for smaller-scale developments) can help ensure that proposals are well integrated with existing facilities and places.

4.4.3   Density is also an important consideration in reducing people's reliance on the private car. PPS3[5] encourages a flexible approach to density, reflecting the desirability of using land efficiently, linked to the impacts of climate change. It sets a national minimum indicative density of 30 dwellings per hectare. Residential densities should be planned to take advantage of a proximity to activities, or to good public transport linking those activities. *Better Places to Live: By Design*[6] advises that a certain

critical mass of development is needed to justify a regular bus service, at frequent intervals, which is sufficient to provide a real alternative to the car.

4.5 Layout considerations

4.5.1   Streets are the focus of movement in a neighbourhood. Pedestrians and cyclists should generally share streets with motor vehicles. There will be situations where it is appropriate to include routes for pedestrians and cyclists segregated from motor traffic, but they should be short, well overlooked and relatively wide to avoid any sense of confinement. It is difficult to design an underpass or alleyway which satisfies the requirement that pedestrians or cyclists will feel safe using them at all times.

4.5.2   The principle of integrated access and movement means that the perimeter block is usually an effective structure for residential neighbourhoods. A block structure works in terms of providing direct, convenient, populated and overlooked routes. In addition, it makes efficient use of land, offers opportunities for enclosed private or communal gardens, and is a tried and tested way of creating quality places (Figs 4.6 and 4.7).

4.5.3   Several disadvantages have become apparent with housing developments built in the last 40 years which departed from traditional arrangements. Many have layouts that make orientation difficult, create left-over or ill-defined spaces, and have too many blank walls or façades. They can also be inconvenient for pedestrians, cyclists and bus users.

4   DETR (2001) *Policy Planning Guidance 13: Transport.* London: TSO.
5   DTLR and CABE (2001) *Better Places to Live: By Design. A Companion Guide to PPG3.* London: Thomas Telford Ltd.
6   Communities and Local Government (2006) *Planning Policy Statement 3: Housing.* London: TSO.

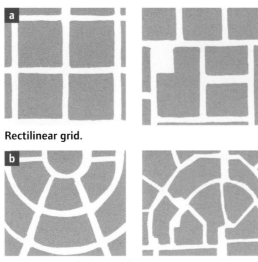

Rectilinear grid.

Concentric grids designed to promote access to local centres or public transport routes.

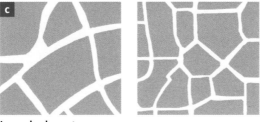

Irregular layouts.

Figure 4.8 Variations on the block structure.

4.5.4   Within a block structure, the designer has more freedom to create innovative layouts. The layouts in Fig. 4.8, and variations on them (such as a 'broken grid' with the occasional cul-de-sac), are recommended when planning residential and mixed-use neighbourhoods.

Geometric choices and street pattern

4.5.5   Straight streets are efficient in the use of land. They maximise connections between places and can better serve the needs of pedestrians who prefer direct routes. However, long, straight streets can also lead to higher speeds. Short and curved or irregular streets contribute to variety and a sense of place, and may also be appropriate where there are topographical or other site constraints, or where there is a need to introduce some variation for the sake of interest. However, layouts that use excessive or gratuitous curves should be avoided, as they are less efficient and make access for pedestrians and cyclists more difficult.

4.5.6   Geometric choices and street pattern should be based on a thorough understanding of context.

Figure 4.9 A good example of a pedestrian/cycle route at Poundbury, Dorchester. It is short, direct and with good surveillance.

4.5.7   Cul-de-sacs may be required because of topography, boundary or other constraints. Cul-de-sacs can also be useful in keeping motor-traffic levels low in a particular area, but any through connections for pedestrians and cyclists should be well overlooked with active frontages. Cul-de-sacs can also provide the best solution for developing awkward sites where through routes are not practical (Fig. 4.9). Caution must, however, be exercised when planning for cul-de-sacs, as they may concentrate traffic impact on a small number of dwellings, require turning heads that are wasteful in land terms and lead to additional vehicle travel and emissions, particularly by service vehicles.

4.6   Crime prevention

4.6.1   The layout of a residential area can have a significant impact on crime against property (homes and cars) and pedestrians. Section 17 of the Crime and Disorder Act 1998,[7] requires local authorities to exercise their function with due regard to the likely effect on crime and disorder. To ensure that crime prevention considerations are taken into account in the design of layouts, it is important to consult police architectural liaison officers and crime prevention officers, as advised in *Safer Places*.[8]

4.6.2   To ensure that crime prevention is properly taken into account, it is important that the way in which permeability is provided is given careful consideration. High permeability is conducive to walking and cycling, but can lead to problems of anti-social behaviour if it is only achieved by providing routes that are poorly overlooked, such as rear alleyways.

7   Crime and Disorder Act 1998. London: TSO.
8   ODPM and Home Office (2004) *Safer Places: The Planning System and Crime Prevention*. London: Thomas Telford Ltd.

Manual for Streets

4.6.3   *Safer Places* highlights the following principles for reducing the likelihood of crime in residential areas (*Wales*: also refer to Technical Advice Note (TAN) 12[9]):

- the desire for connectivity should not compromise the ability of householders to exert ownership over private or communal 'defensible space';
- access to the rear of dwellings from public spaces, including alleys, should be avoided – a block layout, with gardens in the middle, is a good way of ensuring this;
- cars, cyclists and pedestrians should be kept together if the route is over any significant length – there should be a presumption against routes serving only pedestrians and/or cyclists away from the road unless they are wide, open, short and overlooked;
- routes should lead directly to where people want to go;
- all routes should be necessary, serving a defined function;
- cars are less prone to damage or theft if parked in-curtilage (but see Chapter 8). If cars cannot be parked in-curtilage, they should ideally be parked on the street in view of the home. Where parking courts are used, they should be small and have natural surveillance;
- layouts should be designed with regard to existing levels of crime in an area; and
- layouts should provide natural surveillance by ensuring streets are overlooked and well used (Fig. 4.10).

9   Welsh Assembly Government (2002). *Technical Advice Note 12: Design*. Cardiff: NAfW. Chapter 5, Design Issues.

Figure 4.10 Active frontage to all streets and to neighbouring open space should be an aim in all developments. Blank walls can be avoided, even on the return at junctions, with specially designed house types.

4.7   Street character types

4.7.1   Traditionally, road hierarchies (e.g. district distributor, local distributor, access road, etc.) have been based on traffic capacity. As set out in Chapter 2, street character types in new residential developments should be determined by the relative importance of both their place and movement functions.

4.7.2   Examples of the more descriptive terminology that should now be used to define street character types are
· high street;
· main street;
· shopping street;
· mixed-use street;
· avenue;
· boulevard;
· mews;
· lane;
· courtyard;

4.7.3   The above list is not exhaustive. Whatever terms are used, it is important that the street character type is well defined, whether in a design code or in some other way. The difference in approach is illustrated by Figs 4.11 and 4.12.

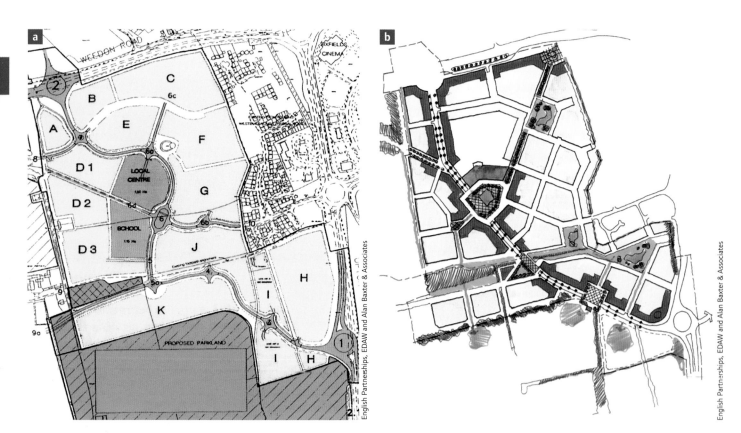

Figure 4.11 Alternative proposals for a development: (a) is highways-led; while (b) is more attuned to pedestrian activity and a sense of place.

Figure 4.12 (a) Existing development in Upton turns its back on the street; while (b) a later development has a strong presence on the street. The latter was delivered using a collaborative workshop design process and a design code.

5

Quality places

Chapter aims

- Promote the place function of streets and explain the role that streets can play in making better places.

- Stress the importance and value of urban design as a framework within which streets are set out and detailed.

- Set out expectations for the design of quality places, as well as routes for safe and convenient movement.

- Discuss local distinctiveness.

5.1 Introduction

5.1.1 The previous chapter described how to plan sustainable communities, covering issues such as the need to plan for connected layouts, mixed uses and walkable neighbourhoods. This chapter develops those themes by demonstrating the importance of quality and encouraging the use of three-dimensional urban design.

5.2 The value of good design

5.2.1 Good design plays a vital role in securing places that are socially, economically and environmentally sustainable (see 'Gateshead case study box'). Planning Policy Statement 1: Delivering Sustainable Development (PPS1)[1] emphasises this. It states that 'good design ensures attractive, usable, durable and adaptable places and is a key element in achieving sustainable development. Good design is indivisible from good planning ... and should contribute positively to making places better for people' (*Wales*: refer to *Planning Policy Wales*,[2] Section 2.9, and *Technical Advice Note (TAN) 12*[3]).

5.2.2 This message is also reinforced by *Planning Policy Statement 3: Housing* (PPS3)[4] which states that 'good design is fundamental to the development of high-quality new housing, which contributes to the creation of sustainable, mixed communities'. (*Wales*: refer to *Ministerial Interim Planning Policy Statement 01/2006: Housing*[5]).

5.2.3 There is growing evidence of the benefits of a public space, development or building that improves people's sense of well being, although these benefits can often be difficult to quantify.

1 ODPM (2005) *Planning Policy Statement 1: Delivering Sustainable Developments*. London: TSO.
2 Welsh Assembly Government (2002) *Planning Policy Wales*. Cardiff: NAfW.
3 Welsh Assembly Government (2002) *Technical Advice Note 12: Design*. Cardiff: NAfW.
4 Communities and Local Government (2006) *Planning Policy Statement 3: Housing*. London: TSO.
5 Welsh Assembly Government (2002) *Ministerial Interim Planning Policy Statement 01/2006: Housing*. Cardiff: NAfW.
6 CABE (2002) *The Value of Good Design*. London CABE; CABE (2006) *Buildings and Spaces: Why Design Matters*. London: CABE; CABE (2006) *The Value Handbook*. London: CABE; and CABE (2006) *The Cost of Bad Design*. London: CABE.

Staiths South Bank, Gateshead

Figure 5.1 New development at Staiths South Bank, Gateshead.

- A significant level of detailed effort was required to negotiate deviation from standards – this was resource intensive. MfS guidance aims to avoid this by promoting the acceptance of innovation (Fig. 5.1).
- The homes are relatively affordable which shows that high-quality design need not be expensive.
- Parking was limited to a ratio of one space per house, which provided scope for a higher-quality public realm.
- The scheme was designed as a Home Zone.

However, evidence is also growing of the economic, social and environmental benefits of good urban design. Good design should not be considered as an optional or additional expense – design costs are only a small percentage of construction costs, but it is through the design process that the largest impact can be made on the quality, efficiency and overall sustainability of buildings, and on the long-term costs of maintenance and management (Fig. 5.2).

5.2.4 CABE has collated a supporting evidence base,[6] which includes the following:
- compact neighbourhoods that integrate parking and transport infrastructure, encourage walking and cycling, and so reduce fuel consumption;
- properties adjacent to a good-quality park have a 5–7% price premium compared with identical properties in the same area but that are away from the park; and
- the benefits of better-designed commercial developments include higher rent levels, lower maintenance costs, enhanced regeneration and increased public support for the development.

Figure 5.2 Newhall, Harlow – a masterplan-led approach with bespoke housing design.

5.3 Key aspects of urban design

'Urban design is the art of making places for people. It includes the way places work and matters such as community safety, as well as how they look. It concerns the connections between people and places, movement and urban form, nature and the built fabric, and the processes for ensuring successful villages, towns and cities.'

By Design: Urban Design in the Planning System: Towards Better Practice[7]

5.3.1   It is important to appreciate what this means in practice. It is easy to advocate places of beauty and distinct identity, but it takes skill to realise them and ensure they are fit for purpose. A number of key documents and initiatives provide an introduction, including the *Urban Design Compendium*,[8] *Better Places to Live: By Design*[9] and *Building for Life*[10] (see box) (*Wales*: see also *Creating Sustainable Places*[11] and *A Model Design Guide for Wales*[12]).

5.3.2   These basic aspects of urban design, however, are not being realised in many new developments. All too often, new development lacks identity and a sense of place. In these cases, it lets communities and users down, and undermines the aims of the sustainable communities agenda.

5.3.3   Frequently, it is in the interaction between the design and layout of homes and streets that attempts to create quality places break down.[13] In the past, urban designers sometimes felt that their schemes were compromised by the application of geometrical standards to highways that were current at the time. Highway engineers, in turn, have occasionally raised concerns about layouts that did not comply with the design criteria to which they were working.

5.3.4   MfS advocates better co-operation between disciplines, and an approach to design based on multiple objectives.

5.4 Street dimensions

5.4.1   Most neighbourhoods include a range of street character types, each with differing characteristics, including type of use, width and building heights. These characteristics dictate how pedestrians and traffic use the street.

Width

5.4.2   Width between buildings is a key dimension and needs to be considered in relation to function and aesthetics. Figure 5.3 shows typical widths for different types of street. The distance between frontages in residential streets typically ranges from 12 m to 18 m, although there are examples of widths less than this working well. There are no fixed rules but account should be taken of the variety of activities taking place in the street and of the scale of the buildings on either side.

7   DETR and CABE (2000) By Design: *Urban Design in the Planning System: Towards Better Practice*. London: Thomas Telford Ltd.

8   Llewelyn Davies (2000) *The Urban Design Compendium*. London: English Partnerships and The Housing Corporation.

9   DTLR and CABE (2001) *Better Places to Live: By Design. A Companion Guide to PPG3*. London: Thomas Telford Ltd.

10 www.buildingforlife.org.uk.

11 Welsh Development Agency (WDA) (2005) *Creating Sustainable Places*. Cardiff: WDA.

12 LDA Design (2005) *A Model Design Guide for Wales: Residential Development*. Cardiff: Planning Officers Society Wales.

13 CABE (2005) *Housing Audit: Assessing the Design Quality of New Homes in the North East, North West and Yorkshire & Humber*. London: Ernest Bond Printing Ltd.

Manual for Streets

The principles of urban design

The fundamental principles of urban design are described more fully in *By Design: Urban Design in the Planning System: Towards Better Practice*.[14] They involve expressing the main objectives of urban design through the various aspects of the built form.

The objectives of urban design can be summarised as follows:

- *Character* – a place with its own identity.
- *Continuity and enclosure* – a place where public and private spaces are clearly distinguished.
- *Quality of the public realm* – a place with attractive and successful outdoor areas.
- *Ease of movement* – a place that is easy to get to and move through.
- *Legibility* – a place that has a clear image and is easy to understand.
- *Adaptability* – a place that can change easily.
- *Diversity* – a place with variety and choice.

The aspects of the built form are described as follows:

- *Layout: urban structure* – the framework of routes and spaces that connect locally and more widely, and the way developments, routes and open spaces relate to one another.
- *Layout: urban grain* – the pattern of the arrangement of street blocks, plots and their buildings in a settlement.
- *Landscape* – the character and appearance of land, including its shape, form, ecology, natural features, colours and elements, and the way these components combine.
- *Density and mix* – the amount of development on a given piece of land and the range of uses. Density influences the intensity of development, and, in combination with the mix of uses, can affect a place's vitality and viability.
- *Scale: height* – scale is the size of a building in relation to its surroundings, or the size of parts of a building or its details, particularly in relation to the size of a person. Height determines the impact of development on views, vistas and skylines.
- *Scale: massing* – the combined effect of the arrangement, volume and shape of a building or group of buildings in relation to other buildings and spaces.
- *Appearance: details* – the craftsmanship, building techniques, decoration, styles and lighting of a building or structure.
- *Appearance: materials* – the texture, colour, pattern and durability of materials, and how they are used.

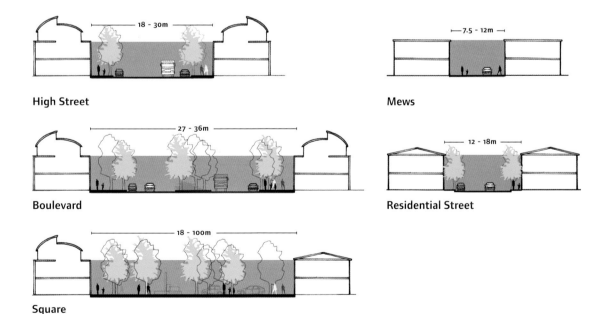

High Street

Boulevard

Square

Mews

Residential Street

Figure 5.3 Typical widths for different types of street.

14 DETR/CABE (2000) *By Design: Urban Design in the Planning System: Towards Better Practice*. London: Thomas Telford.

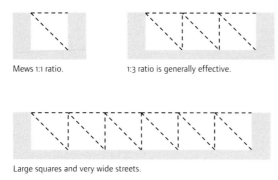

Mews 1:1 ratio.

1:3 ratio is generally effective.

Spatial definition of street through use of planting.

Large squares and very wide streets.

Spatial definition by building height.

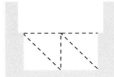

Spatial definition by recess line.

Figure 5.4 Height-to-width ratios.

Height

5.4.3   The public realm is defined by height as well as width – or, more accurately, the ratio of height to width. It is therefore recommended that the height of buildings (or mature trees where present in wider streets) is in proportion to the width of the intervening public space to achieve enclosure. The actual ratio depends on the type of street or open space being designed for. This is a fundamental urban design principle. The height-to-width enclosure ratios shown in Table 5.1 and illustrated in Fig. 5.4 can serve as a guide.

Table 5.1 Height-to-width ratios

	Maximum	Minimum
Minor streets, e.g. mews	1:1.5	1:1
Typical streets	1:3	1:1.5
Squares	1:6	1:4

5.4.4   The benefits of taller buildings, such as signifying locations of visual importance, adding variety, or simply accommodating larger numbers of dwellings, must be weighed against the possible disadvantages. These include an overbearing relationship with the street, overshadowing of surrounding areas, and the need to provide more parking. Design mitigation techniques, such as wider footways, building recesses and street trees, can reduce the impact of taller buildings on their settings (Fig. 5.5).

Length

5.4.5   Street length can have a significant effect on the quality of a place. Acknowledging and framing vistas and landmarks can help bring an identity to a neighbourhood and orientate users. However, long straights can encourage high traffic speeds, which should be mitigated through careful design (see Section 7.4 'Achieving appropriate traffic speeds').

5.5   Buildings at junctions

5.5.1   The arrangement of buildings and footways has a major influence on defining the space at a junction. It is better to design the junction on this basis rather than purely on vehicle movement (Fig. 5.6). In terms of streetscape, a wide carriageway with tight, enclosed corners makes a better junction than cutback corners with a sweeping curve. This might involve bringing buildings forward to the corner. Double-fronted buildings also have an important role at corners. Junction treatments are explored in more detail in Chapter 7.

Figure 5.5 Two streets demonstrating different levels of enclosure. Street (a) has a height-to-width ratio of approximately 1:3, enabling a pleasant living environment to be shared with functionality in the form of traffic movement and on-street parking, some of it angled. Street (b) has a height-to-width ratio of about 1:1.5. Again, this works well in urban design terms, but the need to accommodate on-street parking has meant that traffic is restricted to one-way movement.

Figure 5.6 Wide, curved junctions reduce enclosure. In this example, the relationship between the buildings and the amenity space at the centre of the circus is diminished.

5.6 Backs and fronts

5.6.1   In general, it is recommended that streets are designed with the backs and fronts of houses and other buildings being treated differently. The basic tenet is 'public fronts and private backs'. Ideally, and certainly in terms of crime prevention, back gardens should adjoin other back gardens or a secure communal space. Front doors should open onto front gardens, small areas in front of the property, or streets.

5.6.2   The desirability of public fronts and private backs applies equally to streets with higher levels of traffic, such as those linking or providing access to residential areas. If such streets are bounded by back-garden fences or hedges, security problems can increase, drivers may be encouraged to speed, land is inefficiently used, and there is a lack of a sense of place (Fig. 5.7). Research carried out for MfS[15] shows that streets with direct frontage access to dwellings can operate safely with significant levels of traffic.

Figure 5.7 (a) and (b) Cul-de-sacs surrounded by a perimeter road that is fronted by back fences – no sense of place, no relationship with its surroundings, no quality, with streets designed purely for vehicles.

15 I. York, A. Bradbury, S. Reid, T. Ewings and R. Paradise (2007) *The Manual for Streets: Redefining Residential Street Design*. TRL Report No. 661. Crowthorne: TRL.

5.7 Designing streets as social spaces

5.7.1   The public realm should be designed to encourage the activities intended to take place within it. Streets should be designed to accommodate a range of users, create visual interest and amenity, and encourage social interaction. The place function of streets may equal or outweigh the movement function, as described in Chapter 2. This can be satisfied by providing a mix of streets of various dimensions, squares and courtyards, with associated 'pocket parks', play spaces, resting places and shelter. The key is to think carefully about the range of desirable activities for the environment being created, and to vary designs to suit each place in the network.

5.7.2   High-quality open space is a key component of successful neighbourhoods. Local Development Frameworks, often supplemented by open space strategies and public realm strategies, should set out the requirements for provision in particular localities. As with streets, parks and other open spaces should be accessible and be well overlooked[16] (Wales: Refer to TAN 16[17]). Open spaces can aid urban cooling to help mitigate the effects of climate change.

5.8 Other layout considerations

5.8.1   The layout of a new housing or mixed-use area will need to take account of factors other than street design and traffic provision. They include:
- the potential impact on climate change, such as the extent to which layouts promote sustainable modes of transport or reduce the need to travel;
- climate and prevailing wind, and the impact of this on building type and orientation;
- energy efficiency and the potential for solar gain by orientating buildings appropriately;
- noise pollution, such as from roads or railways;
- providing views and vistas, landmarks, gateways and focal points to emphasise urban structure, hierarchies and connections, as well as variety and visual interest;
- crime prevention, including the provision of defensible private and communal space, and active, overlooked streets (see Chapter 4); and
- balancing the need to provide facilities for young children and teenagers overlooked by housing, with the detrimental effects of noise and nuisance that may result.

Figure 5.8 A contemporary interpretation of the terraced house, providing active frontage to the street and a small private buffer area.

5.8.2   Often satisfying one consideration will make it difficult to satisfy another, and invariably a balance has to be achieved. This is one of the reasons for agreeing design objectives at an early stage in the life of the scheme.

5.9 Where streets meet buildings

5.9.1   The space between the front of the building and the carriageway, footway or other public space needs to be carefully managed as it marks the transition from the public to the private realm. Continuous building lines are preferred as they provide definition to, and enclosure of, the public realm. They also make navigation by blind and partially-sighted people easier.

5.9.2   For occupiers of houses, the amenity value of front gardens tends to be lower when compared to their back gardens and increased parking pressures on streets has meant that many householders have converted their front gardens to hard standing for car parking. However, this is not necessarily the most desirable outcome for street users in terms of amenity and quality of place, and can lead to problems with drainage. Where no front garden is provided, the setback of dwellings from the street is a key consideration in terms of:

16   ODPM (2002) *Planning Policy Guidance 17: Planning for Open Space, Sport and Recreation*. London: TSO.
17   Welsh Assembly Government (2006) *Draft Technical Advice Note 16: Sport, Recreation and Open Space*. Cardiff: NAfW.

Figure 5.9 Trees, bollards, benches and the litter bin have the potential to clutter this residential square, but careful design means that they add to the local amenity.

- defining the character of the street;
- determining a degree of privacy;
- security space, providing a semi-private buffer which intruders would have to pass through, thus reducing opportunities for crime (Fig. 5.8);
- amenity space for plants or seating, etc.; and
- functional space for rubbish bins, external meters or storage, including secure parking for bicycles.

5.9.3   Keeping garages and parking areas level with, or behind, the main building line can be aesthetically beneficial in townscape terms.

5.10   Reducing clutter

5.10.1   Street furniture, signs, bins, bollards, utilities boxes, lighting and other items which tend to accumulate on a footway can clutter the streetscape. Clutter is visually intrusive and has adverse implications for many disabled people. The agencies responsible for such items and those who manage the street should consider ways of reducing their visual impact and impediment to users.

5.10.2   Examples of reducing the impact include:[18]
- mounting streetlights onto buildings, or traffic signals onto lighting columns;
- locating service inspection boxes within buildings or boundary walls;
- specifying the location and orientation of inspection covers in the footway;
- ensuring that household bins and recycling containers can be stored off the footway; and
- designing street furniture to be in keeping with its surroundings (Fig. 5.9).

5.10.3   Where terraced housing or flats are proposed, it can be difficult to find space for storing bins off the footway. In these circumstances, sub-surface or pop-up waste containers may be a practicable solution (Fig. 5.10).

5.11   Local distinctiveness

5.11.1   Local identity and distinctiveness are important design considerations and can be strengthened by:
- relating the layout to neighbouring development (if it satisfies the basics of good urban design);
- involving the community early on in the design process;

18 Joint Committee on Mobility of Blind and Partially Sighted People (JCMBPS) (2002) *Policy Statement on Walking Strategies*. Reading: JCMBPS.

Manual for Streets

Figure 5.10 Sub-surface recycling bins for communal use.

- using local materials (which may also be better environmentally);
- using grain, patterns and form sympathetic to the predominant vernacular styles (Fig. 5.11), or as established in local supplementary planning documents and/or Character Assessment documents;[19]
- retaining historical associations; and
- engaging with utility companies to ensure that the design, quality and setting of their street furniture does not detract from the overall street design, view points and vistas.

5.11.2   Village and Town Design Statements, which are based on enhancing local character and distinctiveness, can also be a useful tool.

5.12   Planting

5.12.1   Space for planting can be integrated into layout and building designs, and, wherever possible, located on private land or buildings (generous balconies, roof gardens, walls) or public land intended for adoption, including the highway.

5.12.2   Planting adds value; it helps to soften the urban street-scene, creates visual and sensory interest, and improves the air quality and microclimate. It can also provide habitats for wildlife. The aromatic qualities or contrasting colours and textures of foliage are of value to all, and can assist the navigation of those with visual impairment. Flowers and fruit trees add seasonal variety.

5.12.3   Planting can provide shade, shelter, privacy, spatial containment and separation. It can also be used to create buffer or security zones, visual barriers, or landmarks or gateway features. Vegetation can be used to limit forward visibility to help reduce vehicle speeds.

19 For region-specific guidance, see English Heritage's *Streets for All* series at www.english-heritage.org.uk.

Figure 5.11 The Orchard, Lechlade – new housing sympathetic to the local context.

Figure 5.12 Mature trees help to structure the space, while buildings are placed to create a sense of enclosure.

5.12.4   Existing trees can occupy a substantial part of a development site and can have a major influence on layout design and use of the site, especially if they are protected by Tree Preservation Orders. Layouts poorly designed in relation to existing trees, or retaining trees of an inappropriate size, species or condition, may be resented by future occupants and create pressure to prune or remove them in the future. To reduce such problems, specialist advice is needed in the design process. An arboriculturalist will help determine whether tree retention can be successfully integrated within the new development, specify protection measures required during construction, and recommend appropriate replacements as necessary (Fig. 5.12).

5.12.5   Sustainable planting will require the provision of:

- healthy growing conditions;
- space to allow growth to maturity with minimal intervention or management;
- species appropriate to a local sense of place and its intended function, and site conditions; and

- well-informed proposals for new planting (or the retention and protection of existing plants) and longer-term maintenance. These proposals should be agreed with the adopting local or highway authority, trust, residents' or community association or management company.

5.13   Standing the test of time

5.13.1   Places need to look good and work well in the long term. Design costs are only a small percentage of the overall costs, but it is the quality of the design that makes the difference in creating places that will stand the test of time. Well-designed places last longer and are easier to maintain, thus the costs of the design element are repaid over time. The specification for materials and maintenance regimes should be written to provide high standards of durability and environmental performance. Maintenance should be straightforward and management regimes should ensure that there are clear lines of responsibility. These themes are covered further in Chapter 11.

C

Detailed design issues

6

Street users' needs

David Millington Photography

Chapter aims

- Promote inclusive design.
- Set out the various requirements of street users.
- Summarise the requirements for various types of motor vehicle.

6.1 Introduction

6.1.1   Street design should be inclusive. Inclusive design means providing for all people regardless of age or ability. There is a general duty for public authorities to promote equality under the Disability Discrimination Act 2005.[1] There is also a specific obligation for those who design, manage and maintain buildings and public spaces to ensure that disabled people play a full part in benefiting from, and shaping, an inclusive built environment.

6.1.2   Poor design can exacerbate the problems of disabled people – good design can minimise them. Consultation with representatives of various user-groups, in particular disabled people, is important for informing the design of streets. Local access officers can also assist here.

6.1.3   Designers should refer to *Inclusive Mobility*,[2] *The Principles of Inclusive Design*[3] and *Guidance on the Use of Tactile Paving Surfaces* (1999)[4] in order to ensure that their designs are inclusive.

6.1.4   If any aspect of a street unavoidably prevents its use by particular user groups, it is important that a suitable alternative is provided. For example, a safe cycling route to school may be inappropriate for experienced cyclist commuters, while a cycle route for commuters in the same transport corridor may be unsafe for use by children. Providing one as an alternative to the other overcomes these problems and ensures that the overall design is inclusive.

6.1.5   This approach is useful as it allows for the provision of a specialised facility where there is considerable demand for it without disadvantaging user groups unable to benefit from it.

6.2 Requirements for pedestrians and cyclists

6.2.1   When designing for pedestrians or cyclists, some requirements are common to both:
- routes should form a coherent network linking trip origins and key destinations, and they should be at a scale appropriate to the users;
- in general, networks should allow people to go where they want, unimpeded by street furniture, footway parking and other obstructions or barriers;
- infrastructure must not only be safe but also be perceived to be safe – this applies to both traffic safety and crime; and
- aesthetics, noise reduction and integration with surrounding areas are important – the environment should be attractive, interesting and free from graffiti and litter, etc.

6.3 Pedestrians

6.3.1   The propensity to walk is influenced not only by distance, but also by the quality of the walking experience. A 20-minute walk alongside a busy highway can seem endless, yet in a rich and stimulating street, such as in a town centre, it can pass without noticing. Residential areas can offer a pleasant walking experience if good quality landscaping, gardens or interesting architecture are present. Sightlines and visibility towards destinations or intermediate points are important for pedestrian way-finding and personal security, and they can help people with cognitive impairment.

6.3.2   Pedestrians may be walking with purpose or engaging in other activities such as play, socialising, shopping or just sitting. For the purposes of this manual, pedestrians include wheelchair users and people pushing wheeled equipment such as prams.

6.3.3   As pedestrians include people of all ages, sizes and abilities, the design of streets needs to satisfy a wide range of requirements. A street design which accommodates the needs of children and disabled people is likely to suit most, if not all, user types.

6.3.4   Not all disability relates to difficulties with mobility. People with sensory or cognitive impairment are often less obviously disabled,

1   Disability Discrimination Act 2005. London: TSO.
2   Department for Transport (2002) *Inclusive Mobility A Guide to Best Practice on Access to Pedestrian and Transport Infrastructure*. London: Department for Transport.
3   CABE (2006) *The Principles of Inclusive Design (They include you)*. London: CABE.
4   DETR (1999) *Guidance on the Use of Tactile Paving Surfaces*. London: TSO.

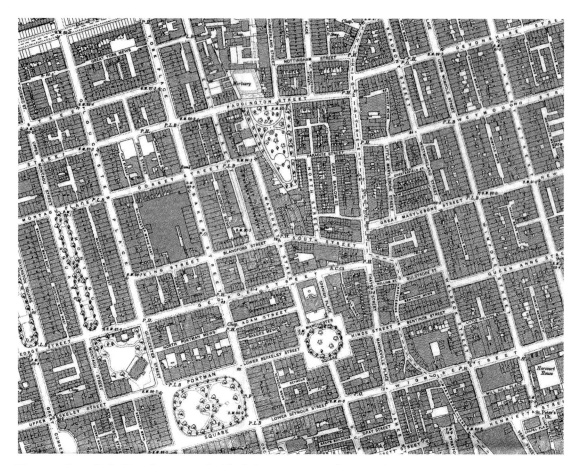

Figure 6.1 West End of London 1884 – the block dimensions are of a scale that encourages walking.

so it is important to ensure that their needs are not overlooked. Legible design, i.e. design which makes it easier for people to work out where they are and where they are going, is especially helpful to disabled people. Not only does it minimise the length of journeys by avoiding wrong turns, for some it may make journeys possible to accomplish in the first place.

6.3.5   The layout of our towns and cities has historically suited pedestrian movement (Fig. 6.1).

6.3.6   Walkable neighbourhoods should be on an appropriate scale, as advised in Chapter 4. Pedestrian routes need to be direct and match desire lines as closely as possible. Permeable networks help minimise walking distances.

6.3.7   Pedestrian networks need to connect with one another. Where these networks are separated by heavily-trafficked roads, appropriate surface level crossings should be provided where practicable. Footbridges and subways should be avoided unless local topography or other conditions make them necessary. The level changes and increased

distances involved are inconvenient, and they can be difficult for disabled people to use. Subways, in particular, can also raise concerns over personal security – if they are unavoidable, designers should aim to make them as short as possible, wide and well lit.

6.3.8   The specific conditions in a street will determine what form of crossing is most relevant. All crossings should be provided with tactile paving. Further advice on the assessment and design of pedestrian crossings is contained in Local Transport Notes 1/95[5] and 2/95[6] and the *Puffin Good Practice Guide*.[7]

6.3.9   Surface level crossings can be of a number of types, as outlined below:
- Uncontrolled crossings – these can be created by dropping kerbs at intervals along a link. As with other types of crossing, these should be matched to the pedestrian desire lines. If the crossing pattern is fairly random and there is an appreciable amount of pedestrian activity, a minimum frequency of 100 m is recommended.[8] Dropped kerbs should

5 Department for Transport (1995) *The Assessment of Pedestrian Crossings*. Local Transport Note 1/95. London: TSO.
6 Department for Transport (1995) *The Design of Pedestrian Crossings*. Local Transport Note 2/95. London: TSO.
7 County Surveyors' Society/Department for Transport (2006) *Puffin Good Practice Guide* available to download from www.dft.gov.uk or www.cssnet.org.uk .
8 Department for Transport (2005) *Inclusive Mobility A Guide to Best Practice on Access to Pedestrian and Transport Infrastructure*. London: Department for Transport.

be marked with appropriate tactile paving and aligned with those on the other side of the carriageway.

- Informal crossings – these can be created through careful use of paving materials and street furniture to indicate a crossing place which encourages slow-moving traffic to give way to pedestrians (Fig. 6.2).
- Pedestrian refuges and kerb build-outs – these can be used separately or in combination. They effectively narrow the carriageway and so reduce the crossing distance. However, they can create pinch-points for cyclists if the remaining gap is still wide enough for motor vehicles to squeeze past them.
- Zebra crossings – of the formal crossing types, these involve the minimum delay for pedestrians when used in the right situation.
- Signalised crossings – there are four types: Pelican, Puffin, Toucan and equestrian crossings. The Pelican crossing was the first to be introduced. Puffin crossings, which have nearside pedestrian signals and a variable crossing time, are replacing Pelican crossings. They use pedestrian detectors to match the length of the crossing period to the time pedestrians take to cross. Toucan and equestrian crossings operate in a similar manner to Puffin crossings except that cyclists can also use Toucan crossings, while equestrian crossings have a separate crossing for horse riders. Signalised crossings are preferred by blind or partially-sighted people.

6.3.10   Obstructions on the footway should be minimised. Street furniture is typically sited on footways and can be a hazard for blind or partially-sighted people.

6.3.11   Where it is necessary to break a road link in order to discourage through traffic, it is recommended that connectivity for pedestrians is maintained through the break unless there are compelling reasons to prevent it.

Figure 6.2 Informal crossing, Colchester – although the chains and a lack of tactile paving are hazardous to blind or partially-sighted people.

Andrew Cameron, WSP

Small radius (eg. 1 metre)

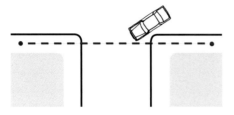

- Pedestrian desire line (---) is maintained.
- Vehicles turn slowly (10 mph – 15 mph).

Large radius (eg. 7 metres)

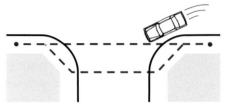

- Pedestrian desire line deflected.
- Detour required to minimise crossing distance.
- Vehicles turn faster (20 mph – 30 mph).

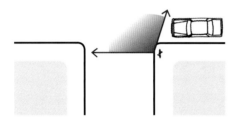

- Pedestrian does not have to look further behind to check for turning vehicles.
- Pedestrian can easily establish priority because vehicles turn slowly.

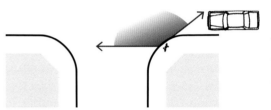

- Pedestrian must look further behind to check for fast turning vehicles.
- Pedestrian cannot normally establish priority against fast turning vehicles.

Figure 6.3 The effects of corner radii on pedestrians.

6.3.12   Pedestrian desire lines should be kept as straight as possible at side-road junctions unless site-specific reasons preclude it. Small corner radii minimise the need for pedestrians to deviate from their desire line (Fig. 6.3). Dropped kerbs with the appropriate tactile paving should be provided at all side-road junctions where the carriageway and footway are at different levels. They should not be placed on curved sections of kerbing because this makes it difficult for blind or partially-sighted people to orientate themselves before crossing.

6.3.13   With small corner radii, large vehicles may need to use the full carriageway width to turn. Swept-path analysis can be used to determine the minimum dimensions required. The footway may need to be strengthened locally in order to allow for larger vehicles occasionally overrunning the corner.

6.3.14   Larger radii can be used without interrupting the pedestrian desire line if the footway is built out at the corners. If larger radii

encourage drivers to make the turn more quickly, speeds will need to be controlled in some way, such as through using a speed table at the junction.

6.3.15   The kerbed separation of footway and carriageway can offer protection to pedestrians, channel surface water, and assist blind or partially-sighted people in finding their way around, but kerbs can also present barriers to some pedestrians. Kerbs also tend to confer an implicit priority to vehicles on the carriageway. At junctions and other locations, such as school or community building entrances, there are benefits in considering bringing the carriageway up flush with the footway to allow people to cross on one level (Fig. 6.4). This can be achieved by:

- raising the carriageway to footway level across the mouths of side roads; and
- providing a full raised speed-table at 'T' junctions and crossroads.

Figure 6.4 Raised crossover, but located away from the desire line for pedestrians and therefore ignored – the crossover should be nearer the junction with, in this case, a steeper ramp for vehicles entering the side street.

Figure 6.6 Uninviting pedestrian link – narrow, not well overlooked, unlit and deserted.

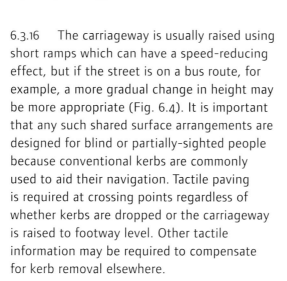

Figure 6.5 Inviting pedestrian link.

Figure 6.7 Overlooked shared route for pedestrians and vehicles, Poundbury, Dorset.

6.3.16   The carriageway is usually raised using short ramps which can have a speed-reducing effect, but if the street is on a bus route, for example, a more gradual change in height may be more appropriate (Fig. 6.4). It is important that any such shared surface arrangements are designed for blind or partially-sighted people because conventional kerbs are commonly used to aid their navigation. Tactile paving is required at crossing points regardless of whether kerbs are dropped or the carriageway is raised to footway level. Other tactile information may be required to compensate for kerb removal elsewhere.

6.3.17   Pedestrians can be intimidated by traffic and can be particularly vulnerable to the fear of crime or anti-social behaviour. In order to encourage and facilitate walking, pedestrians need to feel safe (Figs 6.5 and 6.6).

6.3.18   Pedestrians generally feel safe from crime where:
- their routes are overlooked by buildings with habitable rooms (Fig. 6.7);
- other people are using the street;
- there is no evidence of anti-social activity (e.g. litter, graffiti, vandalised street furniture);
- they cannot be surprised (e.g. at blind corners);
- they cannot be trapped (e.g. people can feel nervous in places with few entry and exit points, such as subway networks); and
- there is good lighting.

6.3.19   Streets with high traffic speeds can make pedestrians feel unsafe. Designers should seek to control vehicle speeds to below 20 mph in residential areas so that pedestrians activity is not displaced. Methods of vehicle speed control are discussed in Chapter 7.

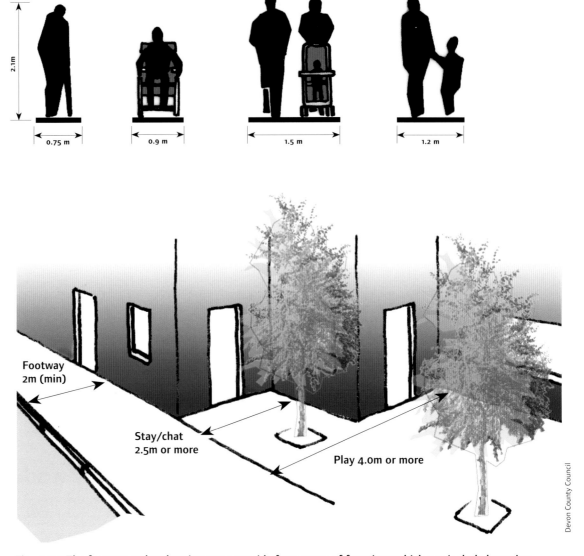

Figure 6.8 The footway and pedestrian areas provide for a range of functions which can include browsing, pausing, socialising and play.

6.3.20   *Inclusive Mobility* gives guidance on design measures for use where there are steep slopes or drops at the rear of footways.

6.3.21   Places for pedestrians may need to serve a variety of purposes, including movement in groups, children's play and other activities (Fig. 6.8).

6.3.22   There is no maximum width for footways. In lightly used streets (such as those with a purely residential function), the minimum unobstructed width for pedestrians should generally be 2 m. Additional width should be considered between the footway and a heavily used carriageway, or adjacent to gathering places, such as schools and shops. Further guidance on minimum footway widths is given in *Inclusive Mobility*.

6.3.23   Footway widths can be varied between different streets to take account of pedestrian volumes and composition. Streets where people walk in groups or near schools or shops, for example, need wider footways. In areas of high pedestrian flow, the quality of the walking experience can deteriorate unless sufficient width is provided. The quality of service goes down as pedestrian flow density increases. Pedestrian congestion through insufficient capacity should be avoided. It is inconvenient and may encourage people to step into the carriageway (Fig. 6.9).

6.3.24   Porch roofs, awnings, garage doors, bay windows, balconies or other building elements should not oversail footways at a height of less than 2.6 m.

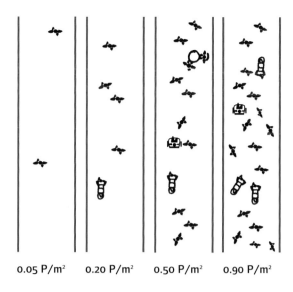

0.05 P/m²   0.20 P/m²   0.50 P/m²   0.90 P/m²

Figure 6.9 Diagram showing different densities of use in terms of pedestrians per square metre. Derived from *Vorrang für Fussgänger* [9].

Figure 6.10 Poorly maintained tree obstructing the footway.

6.3.25   Trees to be sited within or close to footways should be carefully selected so that their spread does not reduce pedestrian space below minimum dimensions for width and headroom (Fig. 6.10).

6.3.26   Low overhanging trees, overgrown shrubs and advertising boards can be particularly hazardous for blind or partially-sighted people. Tapering obstructions, where the clearance under a structure reduces because the structure slopes

down (common under footbridge ramps), or the pedestrian surface ramps up, should be avoided or fenced off.

6.3.27   Designers should attempt to keep pedestrian (and cycle) routes as near to level as possible along their length and width, within the constraints of the site. Longitudinal gradients should ideally be no more than 5%, although topography or other circumstances may make this difficult to achieve (Fig. 6.11).

9   Wissenschaft & Verkehr (1993) *Vorrang für Fussgänger*. Verkehrsclub Österreich.

Figure 6.11 In some instances it may be possible to keep footways level when the carriageway is on a gradient, although this example deflects pedestrians wanting to cross the side road significantly from their desire lines.

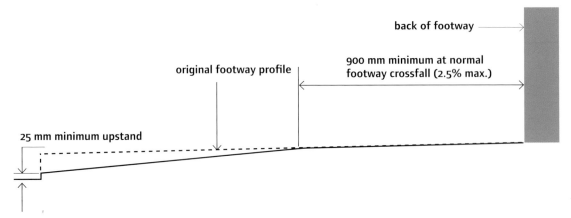

back of footway

900 mm minimum at normal
footway crossfall (2.5% max.)

original footway profile

25 mm minimum upstand

Figure 6.12 Typical vehicle crossover.

6.3.28   Off-street parking often requires motorists to cross footways. Crossovers to private driveways are commonly constructed by ramping up from the carriageway over the whole width of the footway, simply because this is easier to construct. This is poor practice and creates inconvenient cross-falls for pedestrians. Excessive cross-fall causes problems for people pushing prams and can be particularly difficult to negotiate for people with a mobility impairment, including wheelchair users.

6.3.29   Where it is necessary to provide vehicle crossovers, the normal footway cross-fall should be maintained as far as practicable from the back of the footway (900 mm minimum) (Fig. 6.12).

6.3.30   Vehicle crossovers are not suitable as pedestrian crossing points. Blind or partially-sighted people need to be able to distinguish between them and places where it is safe to cross. Vehicle crossovers should therefore have a minimum upstand of 25 mm at the carriageway edge. Where there is a need for a pedestrian crossing point, it should be constructed separately, with tactile paving and kerbs dropped flush with the carriageway.

6.3.31   Surfaces used by pedestrians need to be smooth and free from trip hazards. Irregular surfaces, such as cobbles, are a barrier to some pedestrians and are unlikely to be appropriate for residential areas.

6.3.32   Designs need to ensure that pedestrian areas are properly drained and are neither washed by runoff nor subject to standing water (Fig 6.13).

6.3.33   Seating on key pedestrian routes should be considered every 100 m to provide rest points and to encourage street activity. Seating should ideally be located where there is good natural surveillance.

Figure 6.13 Poor drainage at a pedestrian crossing place causes discomfort and inconvenience.

Figure 6.14 On-street cycling in Ipswich.

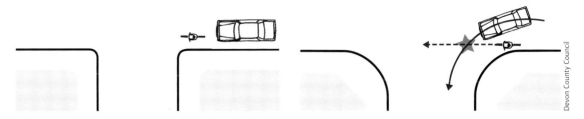

Small radius (eg. 1 metre)

Large radius (eg. 7 metres)

Devon County Council

- Cycle and car speeds compatible.
- Danger from fast turning vehicles cutting across cyclists.

Figure 6.15 The effect of corner radii on cyclists near turning vehicles.

6.4 Cyclists

6.4.1 Cyclists should generally be accommodated on the carriageway. In areas with low traffic volumes and speeds, there should not be any need for dedicated cycle lanes on the street (Fig. 6.14).

6.4.2 Cycle access should always be considered on links between street networks which are not available to motor traffic. If an existing street is closed off, it should generally remain open to pedestrians and cyclists.

6.4.3 Cyclists prefer direct, barrier-free routes with smooth surfaces. Routes should avoid the need for cyclists to dismount.

6.2.4 Cyclists are more likely to choose routes that enable them to keep moving. Routes that take cyclists away from their desire lines and require them to concede priority to side-road traffic are less likely to be used. Anecdotal evidence suggests that cyclists using cycle tracks running adjacent and parallel to a main road are particularly vulnerable when they cross the mouths of side roads and that, overall, these routes can be more hazardous to cyclists than the equivalent on-road route.

6.4.5 Cyclists are particularly sensitive to traffic conditions. High speeds or high volumes of traffic tend to discourage cycling. If traffic conditions are inappropriate for on-street cycling, the factors contributing to them need to be addressed, if practicable, to make on-street cycling satisfactory. This is described in more detail in Chapter 7.

6.4.6 The design of junctions affects the way motorists interact with cyclists. It is recommended that junctions are designed to promote slow motor-vehicle speeds. This may include short corner radii as well as vertical deflections (Fig. 6.15).

6.4.7 Where cycle-specific facilities, such as cycle tracks, are provided, their geometry and visibility should be in accordance with the appropriate design speed. The design speed for a cycle track would normally be 30 km/h (20 mph), but reduced as necessary to as low as 10 km/h (6 mph) for short distances where cyclists would expect to slow down, such as on the approach to a subway. Blind corners are a hazard and should be avoided.

6.4.8 Cyclists should be catered for on the road if at all practicable. If cycle lanes are installed, measures should be taken to prevent them from being blocked by parked vehicles. If cycle tracks are provided, they should be physically segregated from footways/footpaths if there is sufficient width available. However, there is generally little point in segregating a combined width of about 3.3 m or less. The fear of being struck by cyclists is a significant concern for many disabled people. Access officers and consultation groups should be involved in the decision-making process.

6.4.9 Cycle tracks are more suited to leisure routes over relatively open spaces. In a built-up area, they should be well overlooked. The decision to light them depends on the circumstances of the site – lighting may not always be appropriate.

6.4.10   Like pedestrians, cyclists can be vulnerable to personal security concerns. Streets which meet the criteria described for pedestrians are likely to be acceptable to cyclists.

6.4.11   The headroom over routes used by cyclists should normally be 2.7 m (minimum 2.4 m). The maximum gradients should generally be no more than 3%, or 5% maximum over a distance of 100 m or less, and 7% maximum over a distance of 30 m or less. However, topography may dictate the gradients, particularly if the route is in the carriageway.

6.4.12   As a general rule, the geometry, including longitudinal profile, and surfaces employed on carriageways create an acceptable running surface for cyclists. The exception to this rule is the use of granite setts, or similar. These provide an unpleasant cycling experience due to the unevenness of the surface. They can prove to be particularly hazardous in the wet and when cyclists are turning, especially when giving hand signals at the same time. The conditions for cyclists on such surfaces can be improved if the line they usually follow is locally paved using larger slabs to provide a smoother ride.

6.5   Public transport

6.5.1   This section concentrates on bus-based public transport as this is the most likely mode to be used for serving residential areas. *Inclusive Mobility* gives detailed guidance on accessible bus stop layout and design, signing, lighting, and design of accessible bus (and rail) stations and interchanges.

Public transport vehicles

6.5.2   Purpose-built buses, from 'hoppers' to double-deckers, vary in length and height, but width is relatively fixed (Fig. 6.16).

6.5.3   Streets currently or likely to be used by public transport should be identified in the design process, working in partnership with public transport operators.

6.5.4   Bus routes and stops should form key elements of the walkable neighbourhood. Designers and local authorities should try to ensure that development densities will be high enough to support a good level of service without long-term subsidy.

6.5.5   In order to design for long-term viability, the following should be considered:
- streets serving bus routes should be reasonably straight. Straight routes also help passenger demand through reduced journey times and better visibility. Straight streets may, however, lead to excessive speeds. Where it is necessary to introduce traffic-calming features, designers should consider their potential effects on buses and bus passengers; and
- layouts designed with strong connections to the local highway network, and which avoid long one-way loops or long distances without passenger catchments, are likely to be more viable.

6.5.6   Bus priority measures may be appropriate within developments to give more direct routeing or to assist buses in avoiding streets where delays could occur.

6.5.7   Using a residential street as a bus route need not require restrictions on direct vehicular access to housing. Detailed requirements for streets designated as bus routes can be determined in consultation with local public transport operators. Streets on bus routes should not generally be less than 6.0 m wide (although this could be reduced on short sections with good inter-visibility between opposing flows). The presence and arrangement of on-street parking, and the manner of its provision, will affect width requirements.

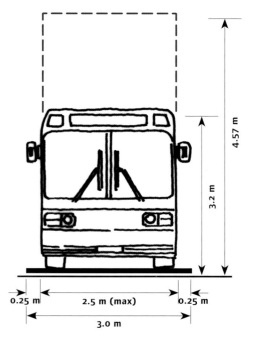

4.57 m

3.2 m

0.25 m   2.5 m (max)   0.25 m

3.0 m

Figure 6.16 Typical bus dimensions

Manual for Streets

Figure 6.17 The bus lay-by facilitates the free movement of other vehicles, but it is inconvenient for pedestrians.

6.5.8   Swept-path analysis can be used to determine the ability of streets to accommodate large vehicles. Bus routes in residential areas are likely to require a more generous swept path to allow efficient operation. While it would be acceptable for the occasional lorry to have to negotiate a particular junction with care, buses need to be able to do so with relative ease. The level of provision required for the movement of buses should consider the frequency and the likelihood of two buses travelling in opposite directions meeting each other on a route.

Bus stops

6.5.9   It is essential to consider the siting of public transport stops and related pedestrian desire lines at an early stage of design. Close co-operation is required between public transport operators, the local authorities and the developer.

6.5.10   First and foremost, the siting of bus stops should be based on trying to ensure they can be easily accessed on foot. Their precise location will depend on other issues, such as the need to avoid noise nuisance, visibility requirements, and the convenience of pedestrians and cyclists. Routes to bus stops must be accessible by disabled people. For example, the bus lay-by in Fig. 6.17 deflects

pedestrians walking along the street from their desire line and the insufficient footway width at the bus stop hinders free movement.

6.5.11   Bus stops should be placed near junctions so that they can be accessed by more than one route on foot, or near specific passenger destinations (schools, shops, etc.) but not so close as to cause problems at the junction. On streets with low movement function (see Chapter 2), setting back bus stops from junctions to maximise traffic capacity should be avoided.

6.5.12   Bus stops should be high-quality places that are safe and comfortable to use. Consideration should be given to providing cycle parking at bus stops with significant catchment areas. Cycle parking should be designed and located so as not to create a hazard, or impede access for, disabled people.

6.5.13   Footways at bus stops should be wide enough for waiting passengers while still allowing for pedestrian movement along the footway. This may require local widening at the stop.

6.5.14   Buses can help to control the speed of traffic at peak times by preventing cars from overtaking. This is also helpful for the safety of passengers crossing after leaving the bus.

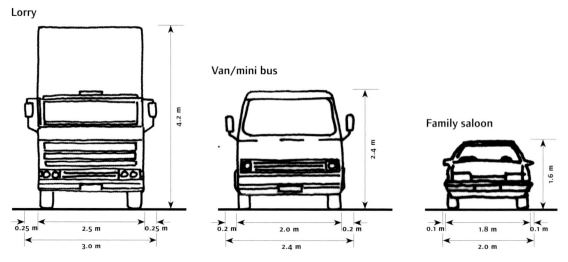

Figure 6.18 Private and commercial motor-vehicles – typical dimensions.

6.6 Private and commercial motor vehicles

6.6.1   Streets need to be designed to accommodate a range of vehicles from private cars, with frequent access requirements, to larger vehicles such as delivery vans and lorries, needing less frequent access (Fig. 6.18). Geometric design which satisfies the access needs of emergency service and waste collection vehicles will also cover the needs of private cars. However, meeting the needs of drivers in residential streets should not be to the detriment of pedestrians, cyclists and public transport users. The aim should be to achieve a harmonious mix of user types.

6.6.2   In a residential environment, flow is unlikely to be high enough to determine street widths, and the extent of parking provision (see Chapter 8) will depend on what is appropriate for the site.

6.6.3   In some locations, a development may be based on car-free principles. For example, there are options for creating developments relatively free of cars by providing remotely sited parking (e.g. Greenwich Millennium Village, see Fig. 6.19) or by creating a wholly car-free development. Such approaches can have a significant effect on the design of residential streets and the way in which they are subsequently used.

Figure 6.19 Greenwich Millennium Village. Cars can be parked on the street for a short time, after which they must be moved to a multi-storey car park.

6.7   Emergency vehicles

6.7.1   The requirements for emergency vehicles are generally dictated by the fire service requirements. Providing access for large fire appliances (including the need to be able to work around them where appropriate) will cater for police vehicles and ambulances.

6.7.2   The Building Regulation requirement B5 (2000)[10] concerns 'Access and Facilities for the Fire Service'. Section 17, 'Vehicle Access', includes the following advice on access from the highway:
- there should be a minimum carriageway width of 3.7 m between kerbs;
- there should be vehicle access for a pump appliance within 45 m of single family houses;
- there should be vehicle access for a pump appliance within 45 m of every dwelling entrance for flats/maisonettes;
- a vehicle access route may be a road or other route; and
- fire service vehicles should not have to reverse more than 20 m.

6.7.3   The Association of Chief Fire Officers has expanded upon and clarified these requirements as follows:
- a 3.7 m carriageway (kerb to kerb) is required for *operating space at the scene of a fire. Simply to reach a fire*, the access route could be reduced to 2.75 m over short distances, provided the pump appliance can get to within 45 m of dwelling entrances;
- if an authority or developer wishes to reduce the running carriageway width to below 3.7 m, they should consult the local Fire Safety Officer;
- the length of cul-de-sacs or the number of dwellings have been used by local authorities as criteria for limiting the size of a development served by a single access route. Authorities have often argued that the larger the site, the more likely it is that a single access could be blocked for whatever reason. The fire services adopt a less numbers-driven approach and consider each application based on a risk assessment for the site, and response time requirements. Since the introduction of the Fire and Rescue Services Act 2004,[11] all regions have had to produce an Integrated Management Plan

setting out response time targets (*Wales: Risk Reduction Plans*[12]). These targets depend on the time required to get fire appliances to a particular area, together with the ease of movement within it. It is therefore possible that a layout acceptable to the Fire and Rescue Service (FRS) in one area, might be objected to in a more remote location;
- parked cars can have a significant influence on response times. Developments should have adequate provision for parking to reduce its impact on response times; and
- residential sprinkler systems are highly regarded by the FRS and their presence allows a longer response time to be used. A site layout which has been rejected on the grounds of accessibility for fire appliances may become acceptable if its buildings are equipped with these systems.

6.8   Service vehicles

6.8.1   The design of local roads should accommodate service vehicles without allowing their requirements to dominate the layout. On streets with low traffic flows and speeds, it may be assumed that they will be able to use the full width of the carriageway to manoeuvre. Larger vehicles which are only expected to use a street infrequently, such as pantechnicons, need not be fully accommodated – designers could assume that they will have to reverse or undertake multi-point turns to turn around for the relatively small number of times they will require access.

6.8.2   Well-connected street networks have significant advantages for service vehicles. A shorter route can be used to cover a given area, and reversing may be avoided altogether. They also minimise land-take by avoiding the need for wasteful turning areas at the ends of cul-de-sacs.

6.8.3   However, some sites cannot facilitate such ease of movement (e.g. linear sites and those with difficult topography), and use cul-de-sacs to make the best use of the land available. For cul-de-sacs longer than 20 m, a turning area should be provided to cater for vehicles that will regularly need to enter the street. Advice on the design of turning areas is given in Chapter 7.

10   Statutory Instrument 2000 No. 2531, The Building Regulations 2000. London: TSO. Part II, paragraph B5: Access and facilities for the fire service.
11   Fire and Rescue Services Act 2004. London: TSO.
12   Risk Reduction Plans required by the Welsh Assembly. See Welsh Assembly Government (2005) *Fire and Rescue National Framework for Wales*. Cardiff: NAfW.

Waste collection vehicles

6.8.4   The need to provide suitable opportunities for the storage and collection of waste is a major consideration in the design of buildings, site layouts and individual streets. Storage may be complicated by the need to provide separate facilities for refuse and the various categories of recyclable waste. Quality of place will be significantly affected by the type of waste collection and management systems used, because they in turn determine the sort of vehicles that will need to gain access.

6.8.5   Policy for local and regional waste planning bodies is set out in *Planning Policy Statement 10: Planning for Sustainable Waste Management* (PPS10)[13] and its companion guide. PPS10 refers to design and layout in new development being able to help secure opportunities for sustainable waste management. Planning authorities should ensure that new developments make sufficient provision for waste management and promote designs and layouts that secure the integration of waste management facilities without adverse impact on the street scene (*Wales*: Refer to Chapter 12 of PPW[14] and TAN 21: Waste[15]).

6.8.6   The operation of waste collection services should be an integral part of street design and achieved in ways that do not compromise quality of place. Waste disposal and collection authorities and their contractors should take into account the geometry of streets across their area and the importance of securing quality of place when designing collection systems and deciding which vehicles are applicable. While it is always possible to design new streets to take the largest vehicle that could be manufactured, this would conflict with the desire to create quality places. It is neither necessary nor desirable to design new streets to accommodate larger waste collection vehicles than can be used within existing streets in the area.

6.8.7   Waste collection vehicles fitted with rear-mounted compaction units (Fig .6.20) are about the largest vehicles that might require regular access to residential areas. BS 5906: 2005[16] notes that the largest waste vehicles currently in use are around 11.6 m long, with

a turning circle of 20.3 m. It recommends a minimum street width of 5 m, but smaller widths are acceptable where on-street parking is discouraged. Swept-path analysis can be used to assess layouts for accessibility. Where achieving these standards would undermine quality of place, alternative vehicle sizes and/or collection methods should be considered.

6.8.8   Reversing causes a disproportionately large number of moving vehicle accidents in the waste/recycling industry. Injuries to collection workers or members of the public by moving collection vehicles are invariably severe or fatal. BS 5906: 2005 recommends a maximum reversing distance of 12 m. Longer distances can be considered, but any reversing routes should be straight and free from obstacles or visual obstructions.

6.8.9   Schedule 1, Part H of the Building Regulations (2000)[17] define locations for the storage and collection of waste. The collection point can be on-street (but see Section 6.8.11), or may be at another location defined by the waste authority. Key points in the Approved Document to Part H are:

·   residents should not be required to carry waste more than 30 m (excluding any vertical distance) to the storage point;

·   waste collection vehicles should be able to get to within 25 m of the storage point (note, BS 5906: 2005[18] recommends shorter distances) and the gradient between the two should not exceed 1:12. There should be a maximum of three steps for waste

Figure 6.20 Large waste collection truck in a residential street.

13 ODPM (2005) *Planning Policy Statement 10: Planning for Sustainable Waste Management*. London: TSO.
14 Welsh Assembly Government (2002). *Planning Policy Wales*. Cardiff: NAfW. Chapter 12, Infrastructure and Services.
15 Welsh Assembly Government (2001) *Technical Advice Note 21: Waste*. Cardiff: NAfW.
16 British Standards Institute (BSI) (2005) *BS 5906: 2005 Waste Management in Buildings – Code of Practice*. London: BSI.
17 Statutory Instrument 2000 No. 2531, The Building Regulations 2000. London: TSO.
18 BSI (2005) *BS 5906: 2005 Waste Management in Buildings – Code of Practice*. London: BSI.

Figure 6.21 Refuse disposal point discharging into underground collection facility.

containers up to 250 litres, and none when larger containers are used (the Health and Safety Executive recommends that, ideally, there should be no steps to negotiate); and

· the collection point should be reasonably accessible for vehicles typically used by the waste collection authority.

6.8.10   Based on these parameters, it may not be necessary for a waste vehicle to enter a cul-de-sac less than around 55 m in length, although this will involve residents and waste collection operatives moving waste the maximum recommended distances, which is not desirable.

6.8.11   BS 5906: 2005 provides guidance and recommendations on good practice. The standard advises on dealing with typical weekly waste and recommends that the distance over which containers are transported by collectors should not normally exceed 15 m for two-wheeled containers, and 10 m for four-wheeled containers.

6.8.12   It is essential that liaison between the designers, the waste, highways, planning and building control authorities, and access officers, takes place at an early stage. Agreement is required on the way waste is to be managed and in particular:

· methods for storing, segregating and collecting waste;
· the amount of waste storage required, based on collection frequency, and the volume and nature of the waste generated by the development; and
· the size of anticipated collection vehicles.

6.8.13   The design of new developments should not require waste bins to be left on the footway as they reduce its effective width. Waste bins on the footway pose a hazard for blind or partially-sighted people and may prevent wheelchair and pushchair users from getting past.

Recycling

6.8.14   The most common types of provision for recycling (often used in combination) are:

· 'bring' facilities, such as bottle and paper banks, where residents leave material for recycling; and
· kerbside collection, where householders separate recyclable material for collection at the kerbside.

6.8.15   'Bring' facilities need to be in accessible locations, such as close to community buildings, but not where noise from bottle banks, etc., can disturb residents. There needs to be enough room for the movement and operation of collection vehicles.

6.8.16   Underground waste containers may be worth considering. All that is visible to the user is a 'litter bin' or other type of disposal point (Fig. 6.21). This collects in underground containers which are emptied by specially equipped vehicles. There were some 175 such systems in use in the UK in 2006.

6.8.17   Kerbside collection systems generally require householders to store more than one type of waste container. This needs to be considered in the design of buildings or external storage facilities.

6.8.18   Designers should ensure that containers can be left out for collection without blocking the footway or presenting hazards to users.

7

Street geometry

Chapter aims

- Advise how the requirements of different users can be accommodated in street design.

- Summarise research which shows that increased visibility encourages higher vehicle speeds.

- Describe how street space can be allocated based on pedestrian need, using swept path analysis to ensure that minimum access requirements for vehicles are met.

- Describe the rationale behind using shorter vehicle stopping distances to determine visibility requirements on links and at junctions.

- Recommend that the design of streets should determine vehicle speed.

- Recommend a maximum design speed of 20 mph for residential streets.

7.1 Introduction

7.1.1 Several issues need to be considered in order to satisfy the various user requirements detailed in Chapter 6, namely:
- street widths and components;
- junctions;
- features for controlling vehicle speeds;
- forward visibility on links; and
- visibility splays at junctions.

7.2 Street dimensions

7.2.1 The design of new streets or the improvement of existing ones should take into account the functions of the street, and the type, density and character of the development.

7.2.2 Carriageway widths should be appropriate for the particular context and uses of the street. Key factors to take into account include:
- the volume of vehicular traffic and pedestrian activity;
- the traffic composition;
- the demarcation, if any, between carriageway and footway (e.g. kerb, street furniture or trees and planting);
- whether parking is to take place in the carriageway and, if so, its distribution, arrangement, the frequency of occupation, and the likely level of parking enforcement (if any);
- the design speed (recommended to be 20 mph or less in residential areas);
- the curvature of the street (bends require greater width to accommodate the swept path of larger vehicles); and
- any intention to include one-way streets, or short stretches of single lane working in two-way streets.

7.2.3 In lightly-trafficked streets, carriageways may be narrowed over short lengths to a single lane as a traffic-calming feature. In such single lane working sections of

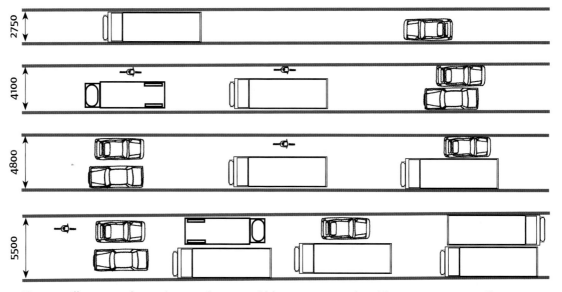

Figure 7.1 Illustrates what various carriageway widths can accommodate. They are not necessarily recommendations.

street, to prevent parking, the width between constraining vertical features such as bollards should be no more than 3.5 m. In particular circumstances this may be reduced to a minimum value of 2.75 m, which will still allow for occasional large vehicles (Fig. 7.1). However, widths between 2.75 m and 3.25 m should be avoided in most cases, since they could result in drivers trying to squeeze past cyclists. The local Fire Safety Officer should be consulted where a carriageway width of less than 3.7 m is proposed (see paragraph 6.6.3)

7.2.4   Each street in the network is allocated a particular street character type, depending on where it sits within the place/movement hierarchy (see Chapter 2) and the requirements of its users (see Chapter 6). Individual streets can then be designed in detail using the relevant typical arrangement as a starting point. For example, one street might have a fairly high movement status combined with a medium place status, whilst another might have very little movement status but a high place status. The typical arrangement for each street character type can then be drawn up. This may be best

Figure 7.3 On-street parking and shallow gradient junction table suitable for accommodating buses.

Newhall demonstrates that adherence to masterplan principles can be achieved through the use of design codes (Fig. 7.3) that are attached to land sales and achieved by covenants.

A list of key dimensions was applied:
· Frontage to frontage – min 10.5 m;
· Carriageway width – min 4.8 m, max 8.8 m;
· Footway width – min 1.5 m;
· Front gardens – min 1.5 m, max 3 m;
· Reservation for services – 1 m; and
· Design speed – 20 mph.

The design is based on pedestrian priority and vehicle speeds of less than 20 mph controlled through the street design.

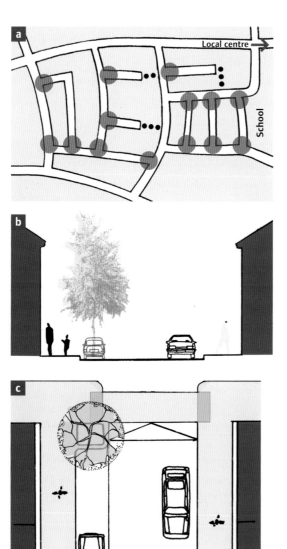

0.3 m 2 m   2 m   4.8 – 5.5 m   2 m 0.3 m

11.4 – 12.1 m

Figure 7.2 Typical representation of a street character type. This example shows the detail for minor side street junctions. Key plan (a) shows the locations, (b) is a cross-section and (c) the plan.

Manual for Streets

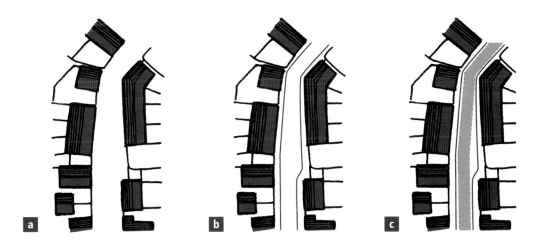

Figure 7.4 Left to right: (a) the buildings and urban edge of a street help to form the place; (b) the kerb line can be used to reinforce this; and (c) the remaining carriageway space is tracked for movement and for the provision of places where people may park their vehicles.

represented using a plan and cross-section as illustrated in Figure 7.2.

7.2.5   These street types can be defined in a design code, as demonstrated at Newhall, Harlow (see Newhall, Harlow box).

Swept path analysis

7.2.6   Swept path analysis, or tracking, is used to determine the space required for various vehicles and is a key tool for designing carriageways for vehicular movement within the overall layout of the street. The potential layouts of buildings and spaces do not have to be dictated by carriageway alignment – they should generally be considered first, with the carriageway alignment being designed to fit within the remaining space (Fig. 7.4).

7.2.7   The use of computer-aided design (CAD) tracking models and similar techniques often proves to be beneficial in determining how the street will operate and how vehicles will move within it. Layouts designed using this approach enable buildings to be laid out to suit the character of the street, with footways and kerbs helping to define and emphasise spaces. Designers have the freedom to vary the space between kerbs or buildings. The kerb line does not need to follow the line of vehicle tracking if careful attention is given to the combination of sightlines, parking and pedestrian movements.

Shared surface streets and squares

7.2.8   In traditional street layouts, footways and carriageways are separated by a kerb. In a street with a shared surface, this demarcation is absent and pedestrians and vehicles share the same surface. Shared surface schemes work best in relatively calm traffic environments. The key aims are to:

- encourage low vehicle speeds;
- create an environment in which pedestrians can walk, or stop and chat, without feeling intimidated by motor traffic;
- make it easier for people to move around; and
- promote social interaction.

7.2.9   In the absence of a formal carriageway, the intention is that motorists entering the area will tend to drive more cautiously and negotiate the right of way with pedestrians on a more conciliatory level (Fig. 7.5).

7.2.10   However, shared surfaces can cause problems for some disabled people. People with cognitive difficulties may find the environment difficult to interpret. In addition, the absence of a conventional kerb poses problems for blind or partially-sighted people, who often rely on this feature to find their way around. It is therefore important that shared surface schemes include an alternative means for visually-impaired people to navigate by.

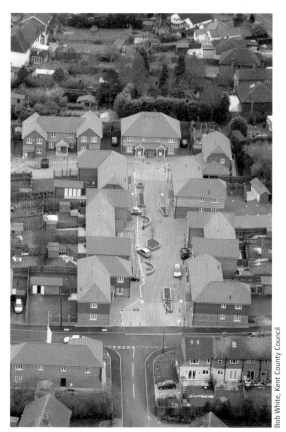

Figure 7.5 A shared surface in a residential area

Bob White, Kent County Council

7.2.11   Research published by the Guide Dogs for the Blind Association in September 2006[1] illustrated the problems that shared surfaces cause for blind or partially-sighted and other disabled people. Further research to be carried out by the Guide Dogs for the Blind Association will consider how the requirements of disabled people can be met, with a view to producing design guidance in due course.

7.2.12   Consultation with the community and users, particularly with disability groups and access officers, is essential when any shared surface scheme is developed. Early indications are that, in many instances, a protected space, with appropriate physical demarcation, will need to be provided, so that those pedestrians who may be unable or unwilling to negotiate priority with vehicles can use the street safely and comfortably.

7.2.13   When designing shared surface schemes, careful attention to detail is required to avoid other problems, such as:
· undifferentiated surfaces leading to poor parking behaviour;
· vulnerable road users feeling threatened by having no space protected from vehicles; and
· the positioning and quantity of planting, street furniture and other features creating visual clutter.

7.2.14   Subject to making suitable provision for disabled people, shared surface streets are likely to work well:
· in short lengths, or where they form cul-de-sacs (Fig. 7.6);
· where the volume of motor traffic is below 100 vehicles per hour (vph) (peak) (see box); and
· where parking is controlled or it takes place in designated areas.

1   The Guide Dogs for the Blind Association (2006) *Shared Surface Street Design Research Project. The Issues: Report of Focus Groups.* Reading: The Guide Dogs for the Blind Association

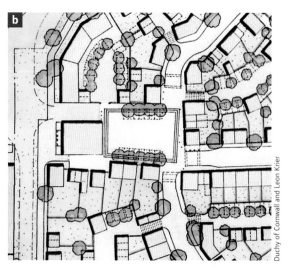

Figure 7.6 (a) and (b) A shared-surface square in Poundbury, Dorset.

Andrew Cameron, WSP

Duchy of Cornwall and Leon Krier

Figure 7.7 A shared surface scheme. Beaulieu Park, Chelmsford.

Figure 7.8 Children playing in a Home Zone, Northmoor, Manchester. However, this type of bollard would cause problems for disabled people.

7.2.15   Shared surface streets are often constructed from paviours rather than asphalt, which helps emphasise their difference from conventional streets. Research for MfS has shown that block paving reduces traffic speeds by between 2.5 and 4.5 mph, compared with speeds on asphalt surfaces (Fig. 7.7).

Home Zones

7.2.16   Home Zones are residential areas designed with streets to be places for people, instead of just for motor traffic. By creating a high-quality street environment, Home Zones strike a better balance between the needs of the local community and drivers (Fig. 7.8). Involving the local community is the key to a successful scheme. Good and effective consultation with all sectors of the community, including young people, can help ensure that the design of individual Home Zones meets the needs of the local residents.

7.2.17   Home Zones often include shared surfaces as part of the scheme design and in doing so they too can create difficulties for disabled people. Research commissioned by the Disabled Persons Transport Advisory Committee (DPTAC) on the implications of Home Zones for disabled people, due to be published in 2007, will demonstrate those concerns. Design guidance relating to this research is expected to be published in due course.

7.2.18   Home Zones are encouraged in both the planning and transport policies for new developments and existing streets. They are distinguished from other streets by having signed entry and exit points, which indicate the special nature of the street.

7.2.19   Local traffic authorities in England and Wales were given the powers to designate roads as Home Zones in section 268 of the Transport Act 2000.[2] The legal procedure for creating a

Research on shared space streets

A study of public transport in London Borough Pedestrian Priority Areas (PPAs) undertaken by TRL for the Bus Priority Team at Transport for London concluded that there is a self-limiting factor on pedestrians sharing space with motorists, of around 100 vph. Above this, pedestrians treat the general path taken by motor vehicles as a 'road' to be crossed rather than as a space to occupy. The speed

of vehicles also had a strong influence on how pedestrians used the shared area. Although this research project concentrated on PPAs, it is reasonable to assume that these factors are relevant to other shared space schemes.

The relationship between visibility, highway width and driver speed identified on links was also found to apply at junctions. A full description of the research findings is available in Manual for Streets: redefining residential street design.[3]

2   I York, A Bradbury, S Reid, T Ewings and R Paradise (2007) *The Manual for Streets: redefining residential street design.* TRL Report No. 661. Crowthorne: TRL.
3   Transport Act 2000. London: TSO.

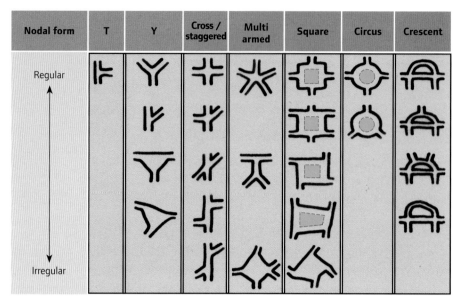

Nodal form	T	Y	Cross / staggered	Multi armed	Square	Circus	Crescent

Fig. 7.9 Illustrative junction layouts.

Home Zone in England is set out in the Quiet Lanes and Home Zones (England) Regulations 2006[4] and guidance is provided in Department for Transport Circular 02/2006.[5] Procedure regulations are yet to be made in Wales, but traffic authorities may still designate roads as Home Zones.

7.2.20   Developers sometimes implement 'Home Zone style' schemes without formal designation. However, it is preferable for the proper steps to be followed to involve the community in deciding how the street will be used.

7.2.21   In existing streets, it is essential that the design of the Home Zone involves significant participation by local residents and local access groups. In new-build situations, a partnership between the developer and the relevant authorities will enable prospective residents to be made aware of the proposed designation of the street as a Home Zone. This will pave the way for the formal consultation procedure once the s treet becomes public highway.

7.2.22   Further guidance on the design of Home Zones is given in *Home Zones: Challenging the Future of Our Streets,*[6] the Institute of Highway Incorporated Engineers' (IHIE) *Home Zone Design Guidelines*[7] and on the website www.homezones.org.uk.

7.3   Junctions

7.3.1   Junctions that are commonly used in residential areas include:

· crossroads and staggered junctions;
· T and Y junctions; and
· roundabouts.

Figure 7.9 illustrates a broader range of junction geometries to show how these basic types can be developed to create distinctive places. Mini-roundabouts and shared surface squares can be incorporated within some of the depicted arrangements.

7.3.2   Junctions are generally places of high accessibility and good natural surveillance. They are therefore ideal places for locating public buildings, shops and public transport stops, etc. Junctions are places of interaction among street users. Their design is therefore critical to achieving a proper balance between their place and movement functions.

7.3.3   The basic junction forms should be determined at the masterplanning stage. At the street design stage, they will have to be considered in more detail in order to determine how they are going to work in practice. Masterplanning and detailed design will cover issues such as traffic priority arrangements, the need, or otherwise, for signs, markings and kerbs, and how property and building lines are related.

4   Statutory Instrument 2006 No. 2082, the Quiet Lanes and home Zones (England) Regulations 2006. London: TSO.
5   Department for Transport (2006) *Circular 02/2006 – The Quiet Lanes and Home Zones (England) Regulations.* London: TSO.
6   Department for Transport (2005) *Home Zones: Challenging the future of our streets.* London: Department for Transport
7   IHIE (2002) *Home Zones Design Guidelines.* London: IHIE

7.3.4   The resulting spaces and townscape should ideally be represented in three dimensions – see box.

7.3.5   Often, the key to a well-designed junction is the way in which buildings are placed around it and how they enclose the space in which the junction sits. Building placement should therefore be decided upon first, with the junction then designed to suit the available space.

7.3.6   Junction design should facilitate direct pedestrian desire lines, and this will often mean using small corner radii. The use of swept path analysis will ensure that the junctions are negotiable by vehicles (Fig. 7.11).

Figure 7.11 Quadrant kerbstones used instead of large radii at junctions reduce the dominance of the carriageway. This is reinforced by the placement and form of the adjacent buildings and the absence of road markings. However, note the lack of dropped kerbs and tactile paving.

Drawing in three dimensions

Presenting design layouts in three dimensions is an important way of looking at aspects of engineering and urban design together (Fig. 7.10). It enables street furniture, lighting, utility equipment and landscaping to be clearly shown. Three-dimensional layouts are also useful in consultation with the public.

Street cross-sections and plans should be developed initially. Perspective or axonometric drawings can then be produced to add clarity and to assist designers in visualising and refining their ideas. Such three-dimensional representation is fairly easy to achieve both by hand and using CAD software. For more complex schemes, a computer-generated 'walk-through' presentation can be used to demonstrate how the proposal will work in practice. It is also a powerful tool for resolving design issues.

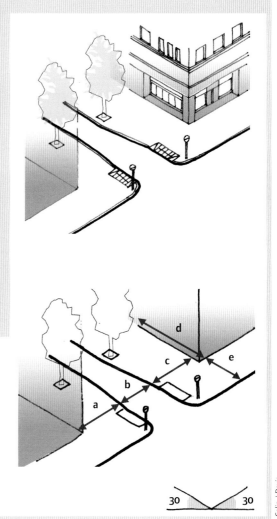

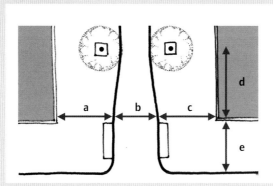

Figure 7.10 Example of three-dimensional presentations.

7.3.7   Junctions can be marked to indicate which arms have priority, but on quieter streets it may be acceptable to leave them unmarked. A lack of marked priority may encourage motorists to slow down to negotiate their way through, making the junction more comfortable for use by pedestrians. However, this approach requires careful consideration (see Chapter 9).

7.3.8   Crossroads are convenient for pedestrians, as they minimise diversion from desire lines when crossing the street. They also make it easier to create permeable and legible street networks.

7.3.9   Permeable layouts can also be achieved using T and Y junctions. Y junctions can increase flexibility in layout design.

7.3.10   Staggered junctions can reduce vehicle conflict compared with crossroads, but may reduce directness for pedestrians. If it is necessary to maintain a view point or vista, and if there is sufficient room between buildings, staggered junctions can be provided within continuous building lines. (Fig. 7.12).

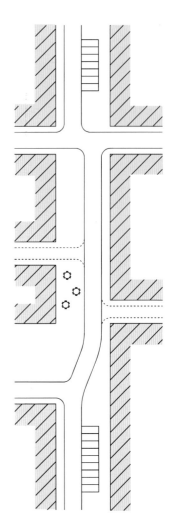

Figure 7.12 – Using staggered junctions to maintain a view point or vista.

Tim Pharoah, Llewelyn Davies Yeang

Case study

Hulme, Manchester: speed tables

Figure 7.13 Raised tables at junction in Hulme. The table has been raised almost to kerb height.

A distinctive feature of the Hulme development is the adherence to a linear grid form. Raised tables at junctions reduce speeds and facilitate pedestrian movement (Fig. 7.13).

7.3.11   Where designers are concerned about potential user conflict, they may consider placing the junction on a speed table (see Hulme, Manchester box). Another option might be to close one of the arms to motor traffic (while leaving it open for pedestrians and cyclists).

7.3.12   Conventional roundabouts are not generally appropriate for residential developments. Their capacity advantages are not usually relevant, they can have a negative impact on vulnerable road users, and they often do little for the street scene.

7.3.13   Larger roundabouts are inconvenient for pedestrians because they are deflected from their desire lines, and people waiting to cross one of the arms may not be able to anticipate easily the movement of motor vehicles on the roundabout, or entering or leaving it.

Figure 7.14 This street avoids the use of vertical traffic-calming features, but the irregular alignment is unsightly and unlikely to have much speed-reducing effect, because of the width of the carriageway. It also results in irregular grassed areas that create a maintenance burden while contributing little to street quality.

7.3.14   Roundabouts can be hazardous for cyclists. Drivers entering at relatively high speed may not notice cyclists on the circulatory carriageway, and cyclists travelling past an arm are vulnerable to being hit by vehicles entering or leaving the junction.

7.3.15   Mini-roundabouts may be more suitable in residential areas, as they cause less deviation for pedestrians and are easier for cyclists to use. In addition, they do not occupy as much land. Practitioners should refer to *Mini-roundabouts: Good Practice Guidelines*.[8]

7.3.16   Continental-style roundabouts are also suitable for consideration. They sit between conventional roundabouts and mini-roundabouts in terms of land take. They retain a conventional central island, but differ in other respects – there is minimal flare at entry and exit, and they have a single-lane circulatory carriageway. In addition, the circulatory carriageway has negative camber, so water drains away from the centre, which simplifies drainage arrangements. Their geometry is effective in reducing entry, circulatory and exit speeds.[9] They are safer for cyclists because of the reduced speeds, together with the fact that drivers cannot overtake on the circulatory carriageway. Their use is described in Traffic Advisory Leaflet 9/97.[10]

Spacing of junctions

7.3.17   The spacing of junctions should be determined by the type and size of urban blocks appropriate for the development. Block size should be based on the need for permeability, and generally tends to become smaller as density and pedestrian activity increases.

7.3.18   Smaller blocks create the need for more frequent junctions. This improves permeability for pedestrians and cyclists, and the impact of motor traffic is dispersed over a wider area. Research in the preparation of MfS[11] demonstrated that more frequent (and hence less busy) junctions need not lead to higher numbers of accidents.

7.3.19   Junctions do not always need to cater for all types of traffic. Some of the arms of a junction may be limited to pedestrian and cycle movement only.

7.4   Achieving appropriate traffic speeds

7.4.1   Conflict among various user groups can be minimised or avoided by reducing the speed and flow of motor vehicles. Ideally, designers should aim to create streets that control vehicle speeds naturally rather than having to rely on unsympathetic traffic-calming measures (Fig. 7.14). In general, providing a separate pedestrian and/or cycle route away from motor traffic should only be considered as a last resort (see the hierarchy of provision in Chapter 4).

8   Department for Transport and County Surveyors' Society (CSS) (2006) *Mini-roundabouts: Good Practice Guidance*. London: CSS.

9   Davies D,G. Taylor, MC, Ryley, TJ, Halliday, M. (1997) *Cyclists at Roundabouts – the Effects of 'Continental' Design on Predicted Safety and Capacity*. TRL Report No. 285. Crowthorne: TRL.

10   DETR (1997) *Traffic Advisory Leaflet 9/97 – Cyclists at Roundabouts: Continental Design Geometry*. London: DETR.

11   I York, A Bradbury, S Reid, T Ewings and R Paradise (2007) *The Manual for Streets: redefining residential street design*. TRL Report no. 661. Crowthorne: TRL.

Andrew Cameron, WSP

Figure 7.15 Trees planted in the highway at Newhall, Harlow, help to reduce vehicle speeds.

7.4.2   For residential streets, a maximum design speed of 20 mph should normally be an objective. The severity of injuries and the likelihood of death resulting from a collision at 20 mph are considerably less than can be expected at 30 mph. In addition, vehicle noise and the intimidation of pedestrians and cyclists are likely to be significantly lower.

7.4.3   Evidence from traffic-calming schemes suggests that speed-controlling features are required at intervals of no more than 70 m in order to achieve speeds of 20 mph or less.[12] Straight and uninterrupted links should therefore be limited to around 70 m to help ensure that the arrangement has a natural traffic-calming effect.

7.4.4   A continuous link can be broken up by introducing features along it to slow traffic. The range of traffic-calming measures available act in different ways, with varying degrees of effectiveness:

- *Physical features* – involving vertical or horizontal deflection – can be very effective in reducing speed. It is preferable to use other means of controlling speeds, if practicable, but there will be situations where physical features represent the optimum solution. Additional sources of advice on traffic calming can be found in Traffic Advisory Leaflet 2/05.[13]
- *Changes in priority* – at roundabouts and other junctions. This can be used to disrupt flow and therefore bring overall speeds down.

- *Street dimensions* – can have a significant influence on speeds. Keeping lengths of street between junctions short is particularly effective. Street width also has an effect on speed (see box).
- *Reduced visibility* – research carried out in preparation of MfS found that reductions in forward visibility are associated with reduced driving speeds (see box).
- *Psychology and perception* – street features and human activity can have an influence on the speed at which people choose to drive. Research[14] suggests that features likely to be effective include the following:
  - edge markings that visually narrow the road – speed reduction is likely to be greatest where the edging is textured to appear unsuitable for driving on;
  - the close proximity of buildings to the road;
  - reduced carriageway width;
  - obstructions in the carriageway (Fig. 7.15);
  - features associated with potential activity in, or close to, the carriageway, such as pedestrian refuges;
  - on-street parking, particularly when the vehicles are parked in echelon formation or perpendicular to the carriageway;
  - the types of land use associated with greater numbers of people, for example shops; and
  - pedestrian activity.

12 DETR (1999) *Traffic Advisory Leaflet 9/99 - 20mph speed limits and zones.* London: DETR.
13 Department for Transport (2005) *Traffic Advisory Leaflet 2/05 - Traffic calming Bibliography.* London: Department for Transport.
14 J Kennedy, R Gorell, L Crinson, A Wheeler, M Elliott (2005) *'Psychological' traffic calming* TRL Report No. 641. Crowthorne: TRL.

Influence of geometry on speed

Research carried out in the preparation of MfS considered the influence of geometry on vehicle speed and casualties in 20 residential and mixed-use areas in the UK. Two highway geometric factors stand out as influencing driving speed, all other things being equal. They are:
· forward visibility; and
· carriageway width.

Improved visibility and/or increased carriageway width were found to correlate with increased vehicle speeds. Increased width for a given visibility, or vice versa, were found to increase speed. These data are summarised in Fig. 7.16.

The relationship between visibility, highway width and driver speed identified on links was also found to apply at junctions. A full description of the research findings is available in TRL Report 661.[15]

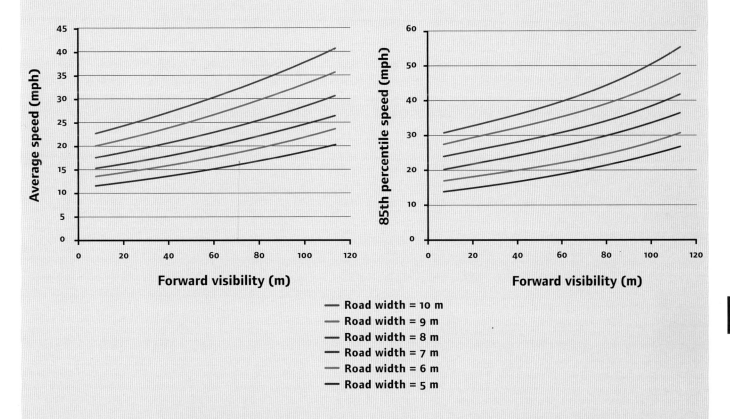

— Road width = 10 m
— Road width = 9 m
— Road width = 8 m
— Road width = 7 m
— Road width = 6 m
— Road width = 5 m

Figure 7.16 Correlation between visibility and carriageway width and vehicle speeds (a) average speeds and (b) 85th percentile speeds. These graphs can be used to give an indication of the speed at which traffic will travel for a given carriageway width/forward visibility combination.

15 I York, A Bradbury, S Reid, T Ewings and R Paradise (2007) *The Manual for Streets: Redefining Residential Street Design*. TRL Report No. 661. Crowthorne: TRL.

7.4.5   Speed limits for residential areas are normally 30 mph, but 20 mph limits are becoming more common. If the road is lit, a 30 mph limit is signed only where it begins – repeater signs are not used here. All other speed limits have to be signed where they start and be accompanied by repeater signs.

7.4.6   A street with a 20 mph limit is not the same as a 20 mph zone. To create a 20 mph zone, it is a legal requirement that traffic-calming measures are installed to ensure that low speeds are maintained throughout. In such cases, the limit is signed only on entering the zone, and no repeater signs are necessary.

7.4.7   Any speed limits below 30 mph, other than 20 mph limits or 20 mph zones, require individual consent from the Secretary of State for Transport. Designers should note that such approval is unlikely to be given.

7.4.8   A speed limit is not an indication of the appropriate speed to drive at. It is the responsibility of drivers to travel within the speed limit at a speed suited to the conditions. However, for new streets, or where existing streets are being modified, and the design speed is below the speed limit, it will be necessary to include measures that reduce traffic speeds accordingly.

7.4.9   Difficulties may be encountered where a new development connects to an existing road. If the junction geometry cannot be made to conform to the requirements for prevailing traffic speeds, the installation of traffic-calming measures on the approach will allow the use of a lower design speed to be used for the new junction.

7.5   Stopping sight distance

7.5.1   This section provides guidance on stopping sight distances (SSDs) for streets where 85th percentile speeds are up to 60 km/h. At speeds above this, the recommended SSDs in the *Design Manual for Roads and Bridges*[16] may be more appropriate.

7.5.2   The stopping sight distance (SSD) is the distance within which drivers need to be able to see ahead and stop from a given speed. It is calculated from the speed of the vehicle, the time required for a driver to identify a hazard and then begin to brake (the perception–reaction time), and the vehicle's rate of deceleration. For new streets, the design speed is set by the designer. For existing streets, the 85th percentile wet-weather speed is used.

7.5.3   The basic formula for calculating SSD (in metres) is:

$$SSD = vt + v^2/2d$$

where:

v = speed (m/s)

t = driver perception–reaction time (seconds)

d = deceleration (m/s^2)

7.5.4   The desirable minimum SSDs used in the *Design Manual for Roads and Bridges* are based on a driver perception–reaction time of 2 seconds and a deceleration rate of 2.45 m/s^2 (equivalent to 0.25g where g is acceleration due to gravity (9.81 m/s^2)). *Design Bulletin 32*[17] adopted these values.

7.5.5   Drivers are normally able to stop much more quickly than this in response to an emergency. The stopping distances given in the Highway Code assume a driver reaction time of 0.67 seconds, and a deceleration rate of 6.57 m/s^2.

7.5.6   While it is not appropriate to design street geometry based on braking in an emergency, there is scope for using lower SSDs than those used in *Design Bulletin 32*. This is based upon the following:

·   a review of practice in other countries has shown that *Design Bulletin 32* values are much more conservative than those used elsewhere;[18]

·   research which shows that the 90th percentile reaction time for drivers confronted with a side-road hazard in a driving simulator is 0.9 seconds (see TRL Report 332[19]);

·   carriageway surfaces are normally able to develop a skidding resistance of at least 0.45g in wet weather conditions. Deceleration rates of 0.25g (the previously assumed value) are more typically associated with snow-covered roads; and

·   of the sites studied in the preparation of this manual, no relationship was found between SSDs and casualties, regardless of whether the sites complied with *Design Bulletin 32* or not.

16   Highways Agency (1992) *Design Manual for Roads and Bridges* London: TSO.

17   Department of the Environment/Department of Transport (1977; 2nd edn 1992) *Design Bulletin 32, Residential Roads and Footpaths - Layout Considerations.* London: HMSO.

18   D.W. Harwood, D.B. Fambro, B. Fishburn, H. Joubert, R. Lamm and B. Psarianos. (1995) *International Sight Distance Design Practices, International Symposium on Highway Geometric Design Practices, Boston, Massachusetts Conference Proceedings.* Washington USA: Transportation Research Board.

19   Maycock G, Brocklebank P. and Hall, R. (1998) *Road Layout Design Standards and Driver Behaviour.* TRL Report No. 332. Crowthorne: TRL

Table 7.1 Derived SSDs for streets (figures rounded).

Speed		16	20	24	25	30	32	40	45	48	50	60
	Kilometres per hour	16	20	24	25	30	32	40	45	48	50	60
	Miles per hour	10	12	15	16	19	20	25	28	30	31	37
SSD (metres)		9	12	15	16	20	22	31	36	40	43	56
SSD adjusted for bonnet length. See 7.6.4		11	14	17	18	23	25	33	39	43	45	59

Additional features will be needed to achieve low speeds

7.5.7 The SSD values used in MfS are based on a perception–reaction time of 1.5 seconds and a deceleration rate of 0.45g (4.41 m/s^2). Table 7.1 uses these values to show the effect of speed on SSD.

7.5.8 Below around 20 m, shorter SSDs themselves will not achieve low vehicle speeds: speed-reducing features will be needed. For higher speed roads, i.e. with an 85th percentile speed over 60 km/h, it may be appropriate to use longer SSDs, as set out in the *Design Manual for Roads and Bridges*.

7.5.9 Gradients affect stopping distances. The deceleration rate of 0.45g used to calculate the figures in Table 7.1 is for a level road. A 10% gradient will increase (or decrease) the rate by around 0.1g.

7.6 Visibility requirements

7.6.1 Visibility should be checked at junctions and along the street. Visibility is measured horizontally and vertically.

7.6.2 Using plan views of proposed layouts, checks for visibility in the horizontal plane ensure that views are not obscured by vertical obstructions.

7.6.3 Checking visibility in the vertical plane is then carried out to ensure that views in the horizontal plane are not compromised by obstructions such as the crest of a hill, or a bridge at a dip in the road ahead. It also takes into account the variation in driver eye height and the height range of obstructions. Eye height is assumed to range from 1.05 m (for car drivers) to 2 m (for lorry drivers). Drivers need to be able to see obstructions 2 m high down to a point 600 mm above the carriageway. The latter dimension is used to ensure small children can be seen (Fig. 7.17).

7.6.4 The SSD figure relates to the position of the driver. However, the distance between the driver and the front of the vehicle is typically up to 2.4 m, which is a significant proportion of shorter stopping distances. It is therefore recommended that an allowance is made by adding 2.4 m to the SSD.

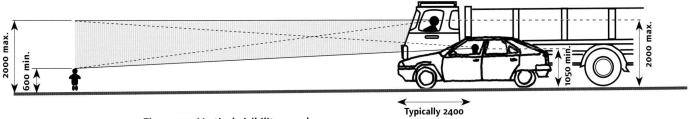

Figure 7.17 Vertical visibility envelope.

7.7 Visibility splays at junctions

7.7.1   The visibility splay at a junction ensures there is adequate inter-visibility between vehicles on the major and minor arms (Fig. 7.18).

7.7.2   The distance back along the minor arm from which visibility is measured is known as the X distance. It is generally measured back from the 'give way' line (or an imaginary 'give way' line if no such markings are provided). This distance is normally measured along the centreline of the minor arm for simplicity, but in some circumstances (for example where there is a wide splitter island on the minor arm) it will be more appropriate to measure it from the actual position of the driver.

7.7.3   The Y distance represents the distance that a driver who is about to exit from the minor arm can see to his left and right along the main alignment. For simplicity it is measured along the nearside kerb line of the main arm, although vehicles will normally be travelling a distance from the kerb line. The measurement is taken from the point where this line intersects the centreline of the minor arm (unless, as above, there is a splitter island in the minor arm).

7.7.4   When the main alignment is curved and the minor arm joins on the outside of a bend, another check is necessary to make sure that an approaching vehicle on the main arm is visible over the whole of the Y distance. This is done by drawing an additional sight line which meets the kerb line at a tangent.

7.7.5   Some circumstances make it unlikely that vehicles approaching from the left on the main arm will cross the centreline of the main arm – opposing flows may be physically segregated at that point, for example. If so, the visibility splay to the left can be measured to the centreline of the main arm.

X distance

7.7.6   An X distance of 2.4 m should normally be used in most built-up situations, as this represents a reasonable maximum distance between the front of the car and the driver's eye.

7.7.7   A minimum figure of 2 m may be considered in some very lightly-trafficked and slow-speed situations, but using this value will mean that the front of some vehicles will protrude slightly into the running carriageway of the major arm. The ability of drivers and cyclists to see this overhang from a reasonable distance, and to manoeuvre around it without undue difficulty, should be considered.

7.7.8   Using an X distance in excess of 2.4 m is not generally required in built-up areas.

7.7.9   Longer X distances enable drivers to look for gaps as they approach the junction. This increases junction capacity for the minor arm, and so may be justified in some circumstances, but it also increases the possibility that drivers on the minor approach will fail to take account of other road users, particularly pedestrians and cyclists. Longer X distances may also result in more shunt accidents on the minor arm. TRL Report No. 184[20] found that accident risk increased with greater minor-road sight distance.

Y distance

7.7.10   The Y distance should be based on values for SSD (Table 7.1).

20 Summersgill I., Kennedy, J. and Baynes, D. (1996) *Accidents at Three-arm Priority Junctions on Urban Single-carriageway Roads* TRL Report no. 184. Crowthorne: TRL.

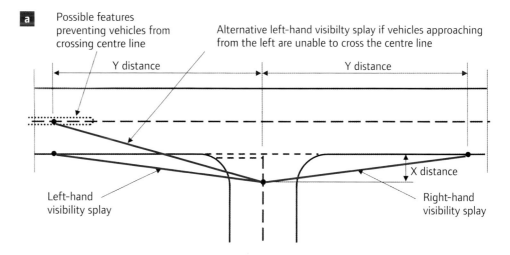

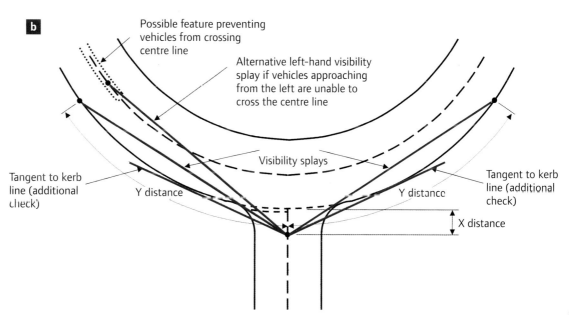

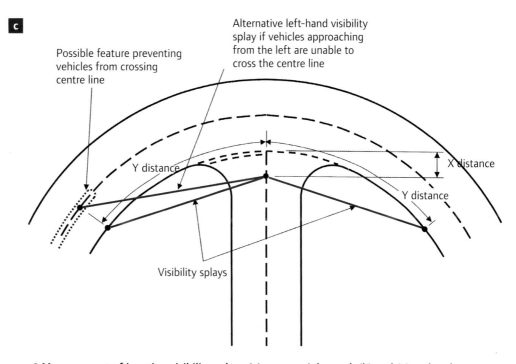

Figure 7.18 Measurement of junction visibility splays (a) on a straight road, (b) and (c) on bends.

7.8   Forward visibility

7.8.1   Forward visibility is the distance a driver needs to see ahead to stop safely for obstructions in the road. The minimum forward visibility required is equal to the minimum SSD. It is checked by measuring between points on a curve along the centreline of the inner traffic lane (see Fig. 7.19).

7.8.2   There will be situations where it is desirable to reduce forward visibility to control traffic speed – the Influence of geometry on speed box describes how forward visibility influences speed. An example is shown in Fig 7.20.

Visibility along the street edge

7.8.3   Vehicle exits at the back edge of the footway mean that emerging drivers will have to take account of people on the footway. The absence of wide visibility splays at private driveways will encourage drivers to emerge more cautiously. Consideration should be given to whether this will be appropriate, taking into account the following:
- the frequency of vehicle movements;
- the amount of pedestrian activity; and
- the width of the footway.

Figure 7.20 Limiting forward visibility helps keep speeds down in Poundbury, Dorset.

7..8.4   When it is judged that footway visibility splays are to be provided , consideration should be given to the best means of achieving this in a manner sympathetic to the visual appearance of the street (Fig. 7.21). This may include:
- the use of boundary railings rather than walls (Fig. 7.22); and
- the omission of boundary walls or fences at the exit location.

Obstacles to visibility

7.8.5   Parking in visibility splays in built-up areas is quite common, yet it does not appear to create significant problems in practice. Ideally, defined parking bays should be provided outside the visibility splay. However, in some circumstances, where speeds are low, some encroachment may be acceptable.

7.8.6   The impact of other obstacles, such as street trees and street lighting columns, should be assessed in terms of their impact on the overall envelope of visibility. In general, occasional obstacles to visibility that are not large enough to fully obscure a whole vehicle or a pedestrian, including a child or wheelchair user, will not have a significant impact on road safety.

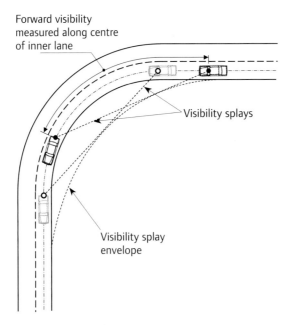

Forward visibility measured along centre of inner lane

Visibility splays

Visibility splay envelope

Figure 7.19 Measurement of forward visibility.

Manual for Streets

Figure 7.21 Beaulieu Park, Chelmsford – low vegetation provides subtle provision of visibility at private driveway.

Figure 7.22 Beaulieu Park, Chelmsford: the visibility splays are provided by railings rather than boundary walls, although the railings could have followed the property boundary.

7.9   Frontage access

7.9.1   One of the key differences between streets and roads is that streets normally provide direct access to buildings and public spaces. This helps to generate activity and a positive relationship between the street and its surroundings. Providing direct access to buildings is also efficient in land-use terms.

7.9.2   The provision of frontage vehicle access onto a street should be considered from the viewpoint of the people passing along the street, as well as those requiring access (Fig. 7.23). Factors to consider include:

- the speed and volume of traffic on the street;
- the possibility of the vehicles turning around within the property – where this is possible, then vehicles can exit travelling forward;
- the presence of gathered accesses – a single access point can serve a number of properties or a communal parking area, for example. This may be acceptable where a series of individual accesses would not be; and

- the distance between the property boundary and the carriageway – to provide adequate visibility for the emerging driver.

7.9.3   In the past, a relatively low limit on traffic flow (300 vehicles per peak hour or some 3,000 vehicles per day) has generally been used when deciding whether direct access was appropriate. This is equivalent to the traffic generated by around 400 houses. Above this level, many local-authority residential road guidelines required the provision of a 'local distributor road'.

Figure 7.23 Frontage access for individual dwellings onto a main street into Dorchester.

7.9.4   Such roads are often very unsuccessful in terms of placemaking and providing for pedestrians and cyclists. In many cases, buildings turn their backs onto local distributors, creating dead frontages and sterile environments. Separate service roads are another possible design response, but these are wasteful of land and reduce visual enclosure and quality.

7.9.5   It is recommended that the limit for providing direct access on roads with a 30 mph speed restriction is raised to at least 10,000 vehicles per day (see box).

Traffic flow and road safety for streets with direct frontage access

The relationship between traffic flow and road safety for streets with direct frontage access was researched for MfS. Data on recorded accidents and traffic flow for a total of 20 sites were obtained. All of the sites were similar in terms of land use (continuous houses with driveways), speed limit (30 mph) and geometry (single-carriageway roads with limited side-road junctions). Traffic flows at the sites varied from some 600 vehicles per day to some 23,000 vehicles per day, with an average traffic flow of some 4,000 vehicles per day.

It was found that very few accidents occurred involving vehicles turning into and out of driveways, even on heavily-trafficked roads.

Links with direct frontage access can be designed for significantly higher traffic flows than have been used in the past, and there is good evidence to raise this figure to 10,000 vehicles per day. It could be increased further, and it is suggested that local authorities review their standards with reference to their own traffic flows and personal injury accident records. The research indicated that a link carrying this volume of traffic, with characteristics similar to those studied, would experience around one driveway-related accident every five years per kilometre. Fewer accidents would be expected on links where the speed of traffic is limited to 20 mph or less, which should be the aim in residential areas.

7.10   Turning areas

7.10.1   Connected street networks will generally eliminate the need for drivers to make three-point turns.

7.10.2   Where it is necessary to provide for three-point turns (e.g. in a cul-de-sac), a tracking assessment should be made to indicate the types of vehicles that may be making this manoeuvre and how they can be accommodated. The turning space provided should relate to its environment, not specifically to vehicle movement (see Fig. 7.24), as this can result in a space with no use other than for turning vehicles. To be effective and usable, the turning head must be kept clear of parked vehicles. Therefore it is essential that adequate parking is provided for residents in suitable locations.

7.10.3   Routeing for waste vehicles should be determined at the concept masterplan or scheme design stage (see paragraph 6.8.4). Wherever possible, routing should be configured so that the refuse collection can be made without the need for the vehicle having to reverse, as turning heads may be obstructed by parked vehicles and reversing refuse vehicles create a risk to other street users.

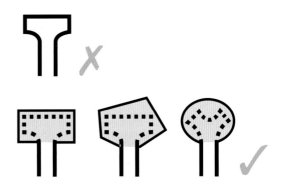

Figure 7.24 Different turning spaces and usable turning heads.

7.11   Overrun areas

7.11.1   Overrun areas are used at bends and junctions (including roundabouts). They are areas of carriageway with a surface texture and/or appearance intended to deter overrunning by cars and other light vehicles. Their purpose is to allow the passage of large vehicles, such as buses and refuse vehicles, while maintaining 'tight' carriageway dimensions that deter smaller vehicles from speeding.

7.11.2   Overrun areas should generally be avoided in residential and mixed-use streets. They can:

· be visually intrusive;
· interfere with pedestrian desire lines (Fig. 7.25); and
· pose a hazard for cyclists.

However, they can help to overcome problems with access for larger vehicles and so may represent the best solution.

Figure 7.25 The overrun area at this junction is hazardous for pedestrians and/or requires them to divert from their desire line. Notice also the unsightly placing of inspection covers. The layout is particularly hazardous for blind and partially-sighted pedestrians.

8

Parking

Chapter aims

- Emphasise the importance of providing sufficient good-quality cycle parking in all new residential developments to meet the needs of residents and visitors.

- Explain how the parking of vehicles is a key function of most streets in residential areas and that it needs to be properly considered in the design process.

- Confirm that, having regard to the policy in *Planning Policy Statement 3: Housing* (PPS3),[1] designers need to consider carefully how to accommodate the number of cars that are likely to be owned by residents (*Wales*: refer to TAN 18: Transport[2]).

- Describe how providing a level of car parking below normal demand levels can be appropriate in some situations.

- Explain the efficiency benefits of unallocated car parking and the need to meet at least some of the normal demand on the street.

- Offer guidance on footway parking.

- Give guidance on the size of parking spaces for cycles, cars and motorcycles.

8.1   Introduction

8.1.1   Accommodating parked vehicles is a key function of most streets, particularly in residential areas. While the greatest demand is for parking cars, there is also a need to consider the parking of cycles, motorcycles and, in some circumstances, service vehicles. Where there is a need to regulate parking, this should be done by making appropriate traffic regulation orders (TROs) and signing and marking in accordance with the Traffic Signs Regulations and General Directions 2002 (TSRGD).[3] Guidance is also provided in the *Traffic Signs Manual*.[4]

8.1.2   The level of parking provision and its location has a key influence on the form and quality of a development, and the choices people make in how they travel. The way cars are

1   Communities and Local Government (2006) *Planning Policy Statement 3: Housing*. London: TSO.
2   Welsh Assembly Government (2007) *Technical Advice Note 18: Transport*. Cardiff: NAfW.
3   Statutory Instrument 2002 No. 3113, The Traffic Signs Regulations and General Directions 2002. London: TSO.
4   Department for Transport (various) *The Traffic Signs Manual*. London: TSO and HMSO.
5   DETR (2001) *Policy Planning Guidance Note 13: Transport*. London: TSO.

parked is a key factor for many issues, such as visual quality, street activity, interaction between residents, and safety.

8.1.3   A failure to properly consider this issue is likely to lead to inappropriate parking behaviour, resulting in poor and unsafe conditions for pedestrians.

8.1.4   Parking can be provided on or off the street. Off-street parking includes parking within a curtilage (on-plot) or in off-street parking areas (off-plot).

8.2   Cycle parking

8.2.1   Providing enough convenient and secure cycle parking at people's homes and other locations for both residents and visitors is critical to increasing the use of cycles. In residential developments, designers should aim to make access to cycle storage at least as convenient as access to car parking.

8.2.2   The need for convenient, safe and secure cycle parking in new developments is recognised in *Policy Planning Guidance Note 13*: Transport (PPG13)[5] (*Wales*: TAN 18), which recommends that provision should be increased to promote cycle use but should at least be at levels consistent with the local authority's cycle target strategy in its Local Transport Plan.

Determining the amount of cycle parking

8.2.3   Shared cycle parking is normally more efficient than providing sufficient space within each dwelling for the maximum possible number of cycles. Shared cycle parking facilities should be secure and convenient to use.

8.2.4   The amount of cycle parking in a shared facility will depend on the overall number of cycles anticipated across the scheme, based on average cycle-ownership levels. This number can vary considerably depending on circumstances.

8.2.5   Houses tend to have higher levels of cycle ownership than flats. Research carried out for CABE/Oxfordshire County Council by WSP

Table 8.1 Average cycle ownership levels in Oxfordshire, 2006

	Average cycles/ dwelling	Average cycles/ resident
Houses, Oxford City	2.65	0.73
Houses, rest of Oxfordshire	1.51	0.52
Flats, Oxford City	0.97	0.48
Flats, rest of Oxfordshire	0.44	0.23

and Phil Jones Associates in 2006 found the average cycle ownership levels shown in Table 8.1.

8.2.6   The amount of provision will also vary depending on the type of development. Cycle use can be expected to be relatively high in places such as student accommodation. In sheltered housing or housing for older people, lower provision is likely to be more appropriate.

8.2.7   When assessing the effect of location, census data on the proportion of trips to work made by cycle provides a useful proxy for assessing the likely level of cycle ownership.

8.2.8   Cycle parking is often likely to be within, or allocated to, individual dwellings, particularly for houses. In such cases, it will be necessary to consider the potential for one cycle to be owned by each resident.

Visitors and mixed-use areas

8.2.9   Providing cycle parking for visitors is important when planning new developments and modifying existing streets. In residential areas, the amount and location of visitor parking can be informed by the amount of cycle parking available to residents and the targeted modal share of visitor trips.

8.2.10   In some cases, visitors may be able to use spare space within residential cycle-parking facilities, whether shared or individual. Some provision in the public realm may also be appropriate, particularly where residents' provision is not easily accessed by visitors.

8.2.11   In mixed-use areas and where there are commercial or communal facilities in a residential neighbourhood, well-located and convenient public cycle-parking will normally be necessary.

Design solutions for residential cycle-parking

8.2.12   Cycles are often kept in garages, and this can be convenient and secure if located near the front of the property. However, garages are not normally designed for cycle storage, and the proportion of housing schemes with individual garages is declining.

8.2.13   Greater consideration therefore needs to be given to the provision of bespoke cycle storage. Cycles are not suited to overnight storage outdoors as they are vulnerable to theft and adverse weather. At the very least, any outdoor cycle parking needs to be covered, and preferably lockable (Fig. 8.1).

8.2.14   If no cycle parking is provided, this may affect the way garages are used. This aspect, among others, will inform decisions on whether garages count fully towards car-parking provision (see paragraph 8.3.41 below).

8.2.15   Where separate cycle-parking is provided within the building, it needs to be conveniently located, close to the main point of access. Where cycle parking is to be provided within a separate building, such as a detached garage or other outbuildings, it will need to be secure, with doors designed for easy access.

8.2.16   In flats, cycle parking has often been inadequate, leading to cycles being stored in hallways or balconies. For new developments, the storage of cycles is an important consideration.

Figure 8.1 Secure cycle storage.

Manual for Streets

Andrew Cameron, WSP

TfL Streetscape Guidance, Transport for London

Figure 8.2 Cycle parking that has good surveillance and is at a key location – in this example near a hospital entrance.

Figure 8.3 Sheffield stands are simple and effective. The design allows the bicycle frame and wheels to be easily locked to the stand. Note the tapping rail near ground level and the reflective bands on the uprights.

8.2.17   For ground-floor flats, or where adequately-sized lifts are provided, storage within the accommodation may be an option, but it will need to be expressly considered in the design and it will be important to ensure that cycles can be brought into the building easily and quickly.

8.2.18   Cycle parking for flats can also be located in communal areas, such as in hallways or under stairs, but, if so, it needs to be properly designed in order to prevent parked cycles becoming a nuisance for residents. If parking is to be located on upper floors, adequately-sized lifts need to be considered.

8.2.19   Another option is to provide communal cycle-parking in secure facilities, such as in underground car parks, in purpose-designed buildings or in extensions to buildings.

8.2.20   Visitor cycle-parking in the public realm is best provided in well-overlooked areas, which may often be the street itself (Fig. 8.2). Although there is a wide variety of design options, simple and unobtrusive solutions, such as Sheffield stands (Fig. 8.3), are preferred. Some bespoke designs are not so convenient, for example they may not allow both wheels to be easily locked to the stand (Fig. 8.4).

8.2.21   Cycle stands need to be located clear of pedestrian desire lines, and generally closer

Cycling England

Figure 8.4 A contemporary design for cycle parking – note that this arrangement is not so convenient for locking both wheels to the stand.

to the carriageway than to buildings. They should be detectable by blind or partially sighted people. A ground level tapping rail at either end of a run of stands should be provided.

8.2.22   The preferred spacing of these stands is about 1 m, so that two cycles can be stored per metre run. Where space is limited, an absolute minimum spacing of 800 mm may be used, although this will make it more difficult for cycles with baskets and panniers to be stored. The outermost stands should be no closer than 550 mm to a parallel wall. In addition, there should be at least 550 mm clear space betwen the ends of individual stands and any wall.

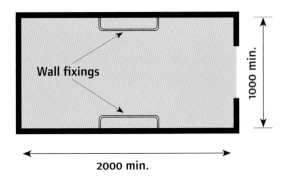

Wall fixings

1000 min.

2000 min.

Figure 8.5 Plan of store for two cycles using wall fixings.

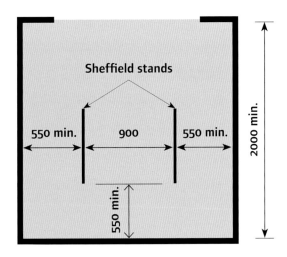

Sheffield stands

550 min.   900   550 min.

2000 min.

550 min.

Figure 8.6 Plan of store for four cycles using Sheffield stands.

8.2.23   Where cycle parking is provided internally, the indicative dimensions shown in Figs 8.5 and 8.6 are appropriate.

8.2.24   Overall space requirements can be reduced where cycles are stored on-end or in two layers using rack systems, but such storage is often not as easy to use by everyone, and is a less desirable option than parking on the ground.

8.3   Car parking

Introduction and policy background

8.3.1   The availability of car parking is a major determinant of travel mode. The Government's general planning policy for car parking is set out in PPG13: Transport. The Government's policy on residential car-parking provision is set out in PPS3: Housing, which is particularly relevant for MfS (*Wales*: policy on parking is set out in *Planning Policy Wales*,[6] supplemented by TAN 18).

8.3.2   PPS3 makes it clear that, when assessing the design quality of a proposed new development, it is important to consider a design-led approach to the provision of car-parking space that is well-integrated with a high-quality public realm. PPS3 (paragraph 51) advises that:

'Local Planning Authorities should, with stakeholders and communities, develop residential parking policies for their areas, taking account of expected levels of car ownership, the importance of promoting good design and the need to use land efficiently.'

8.3.3   The context of a new residential development needs to be carefully considered when determining the appropriate amount of parking (Fig. 8.7). This will be informed by the Transport Assessment, together with any accompanying Travel Plan and the local authority's residential parking policies set out in its Local Development Framework.

8.3.4   Although the ability of residents to reach important destinations by other modes is one factor affecting car ownership, research[7] has shown that dwelling size, type and tenure are also important.

Figure 8.7 Residential car parking.

6   Welsh Assembly Government (2002) *Planning Policy Wales*. Cardiff: NAfW. Chapter 8, Transport.
7   Forthcoming Communities and Local Government research document

Manual for Streets

8.3.5   Local planning authorities will need to consider carefully what is an appropriate level of car parking provision. In particular, under-provision may be unattractive to some potential occupiers and could, over time, result in the conversion of front gardens to parking areas (see box). This can cause significant loss of visual quality and increase rainwater run-off, which works against the need to combat climate change. It is important to be aware that many disabled people are reliant on the use of the private car for personal mobility. Ideally, therefore, layouts should be able to accommodate parking provision for Blue Badge holders.

Car parking provision for new homes

CABE research[8,9] found that car parking remains a significant issue for residents and house buyers. Many people feel that the design for a new residential development should accommodate typical levels of car ownership and that the level of parking in new developments is often inadequate for residents' and visitors' demands. There was a general feeling among buyers of new homes that apparent attempts to restrict parking in order to curb car ownership were unrealistic and had little or no impact on the number of cars a household would require and acquire.

8.3.6   Provision below demand can work successfully when adequate on-street parking controls are present and where it is possible for residents to reach day-to-day destinations, such as jobs, schools and shops, without the use of a car. This will normally be in town and city centres where there will be good public transport and places that can be accessed easily on foot and by cycle. For residents who choose not to own a car, living in such an area may be an attractive proposition.

8.3.7   One way of encouraging reduced car ownership is to provide a car club. Car clubs provide neighbourhood-based short-term car hire to members for periods of as little as one hour, and have been shown to reduce car ownership and use. To function effectively, car club vehicles need to be made available close to members' homes.

8.3.8   More information on car clubs is available at *www.carplus.org.uk* and in the Department for Transport document *Making Car Sharing and Car Clubs Work*[10] (see box).

Car clubs

Making Car Sharing and Car Clubs Work advises that:
'The importance of on-street spaces cannot be underestimated both for open and closed schemes; not least because they provide a very visible image of the presence of a car club, and demonstrate direct benefits for potential users. The provision of dedicated parking spaces is a major incentive for the uptake of community car clubs, particularly in urban areas.'

8.3.9   Highway authorities are able to make TROs, limiting the use of on-street parking spaces to car club vehicles. Authorities that have done this include Bristol, Ealing, Edinburgh, and Kensington and Chelsea. The supporting traffic signs and markings may need to be authorised by the Department for Transport in England or the Welsh Assembly Government (see Fig. 8.8).

Figure 8.8 (a) and (b) A successful car club scheme is operating in Bath, with spaces provided on-street.

8   CABE (2005) *What Home Buyers Want: Attitudes and Decision Making amongst Consumers.* London: CABE.

9   CABE (2005) *What it's Like to Live There: The Views of Residents on the Design of New Housing.* London: CABE.

10   Department for Transport (2004) *Making Car Sharing and Car Clubs Work: A Good Practice Guide.* London: Department for Transport.

Allocated and unallocated parking

8.3.10   Not all parking spaces need to be allocated to individual properties. Unallocated parking provides a common resource for a neighbourhood or a specific development.

8.3.11   A combination of both types of parking can often be the most appropriate solution. There are several advantages to providing a certain amount of unallocated communal parking, and it is recommended that there should be a presumption in favour of including some in most residential layouts. Key considerations for communal parking are that it:

· only needs to provide for average levels of car ownership;
· allows for changes in car ownership between individual dwellings over time;
· provides for both residents' and visitors' needs; and
· can cater for parking demand from non-residential uses in mixed-use areas, which will tend to peak during the daytime when residential demands are lowest.

On-street parking

8.3.12   An arrangement of discrete parking bays adjacent to the running lanes is often the preferred way of providing on-street parking. It has little effect on passing traffic and minimises obstructions to the view of pedestrians crossing the street.

8.3.13   It is recommended that, in most circumstances, at least some parking demand in residential and mixed-use areas is met with well-designed on-street parking (Fig. 8.9).

8.3.14   Breaking up the visual impact can be achieved by limiting on-street parking to small groups of, say, about five spaces. These groups can be separated by kerb build-outs, street furniture or planting.

8.3.15   In planning for expected levels of car ownership it is not always necessary to provide parking on site (i.e. within curtilage or in off-street parking areas). In some cases it may be appropriate to cater for all of the anticipated

Figure 8.9 An example of on-street parking in the centre of the street that helps to separate the car from other users and provides strong surveillance of the cars.

Tim Pharoah, Llewelyn Davies Yeang

Figure 8.10 On-street parking in Crown Street, Glasgow.

demand on-street. This could be the case, for example, with a small infill development where adjacent streets are able to easily accommodate the increase in parking, or where a low car-ownership development is proposed. Crown Street, Glasgow, is an example of a large scheme that has accommodated all parking on-street (Fig. 8.10).

8.3.16   Where regulated on-street parking is provided, it is important to note that it cannot be allocated to individual dwellings, although such spaces can be reserved for particular types of user, such as disabled people.

8.3.17   In deciding how much on-street parking is appropriate, it is recommended that the positive and negative effects listed in the 'On-street parking box' are considered.

On-street parking – positive and negative effects

Positive effects

- A common resource, catering for residents', visitors' and service vehicles in an efficient manner.
- Able to cater for peak demands from various users at different times of the day, for example people at work or residents.
- Adds activity to the street.
- Typically well overlooked, providing improved security.
- Popular and likely to be well-used.
- Can provide a useful buffer between pedestrians and traffic.
- Potentially allows the creation of areas within perimeter blocks that are free of cars.

Negative effects

- Can introduce a road safety problem, particularly if traffic speeds are above 20 mph and there are few places for pedestrians to cross with adequate visibility.
- Can be visually dominant within a street scene and can undermine the established character (Fig. 8.11).
- May lead to footway parking unless the street is properly designed to accommodate parked vehicles.
- Vehicles parked indiscriminately can block vehicular accesses to dwellings.
- Cars parked on-street can be more vulnerable to opportunistic crime than off-street spaces.

Figure 8.11 Street detailing and pedestrian provision dominated by car-parking considerations

8.3.18   Generally the most appropriate solution will be to design for a level of on-street parking that takes account of the following factors:

- the overall level of car ownership in the immediate area;
- the amount of off-street parking provided;
- the amount of allocated parking provided;
- the speed and volume of traffic using the street; and
- the width and geometry of the street and its junctions.

8.3.19   Indicating on-street car-parking spaces clearly through the use of road markings or changes of surfacing material can help to encourage good parking behaviour.

8.3.20   Where on-street spaces are provided in bays adjacent to running lanes, having them drain towards the street will make cleaning easier.

Visitor parking

8.3.21   It is recommended that visitor parking is generally served by unallocated parking, including on-street provision.

8.3.22   Research[11] indicates that no additional provision needs to be made for visitor parking when a significant proportion of the total parking stock for an area is unallocated.

8.3.23   In town centres and other locations with good accessibility by non-car modes, and where on-street parking is controlled, it is often appropriate to omit visitor car-parking spaces.

Car parking provision for disabled people (Blue Badge holders)

8.3.24   Spaces for disabled people[12] need to be properly marked and meet the minimum space requirements (see paragraph 8.3.58 below).

8.3.25   It is preferable to provide these spaces in unallocated areas, including on-street, as it is not normally possible to identify which properties will be occupied by or visited by disabled people. It is recommended that spaces for disabled people are generally located as close as possible to building entrances.

8.3.26   In the absence of any specific local policies, it is recommended that 5% of residential car-parking spaces are designated for use by disabled people. A higher percentage is likely to be necessary where there are proportionally more older residents. Local authorities should provide spaces on the basis of demand.

8.3.27   Where local authorities mark out disabled bays on streets in residential areas, the traffic signs and road markings should comply with TSRGD and be supported by a TRO.

Parking for service vehicles

8.3.28   In most situations, it will not be necessary to provide parking spaces specifically for service vehicles, such as delivery vans, which are normally stationary for a relatively short time. If such parking bays are considered necessary, other vehicles may need to be prevented from using the spaces by regulation and enforcement.

Design and location of car-parking spaces

8.3.29   Guidance on the design and location of car-parking spaces can be found in a number of recent documents.

8.3.30   *Better Places to Live*[13] echoes many of the principles already set out above, including opportunities to use a combination of allocated and unallocated parking and the scope for on-street parking, provided that it is designed so that it is interrupted at regular intervals.

11   Noble, J. and Jenks, M. (1996) *Parking: Demand and Provision in Private Sector Housing Developments.* Oxford: Oxford Brookes University.

12   DETR (2001) *Policy Planning Guidance 13: Transport. London:* TSO. (*Wales:* Welsh Assembly Government (2007) *Technical Advice Note 18:* Transport. Cardiff: NAfW.)

13   DTLR and CABE (2001) *Better Places to Live: By Design. A Companion Guide to PPG3.* London: Thomas Telford Ltd.

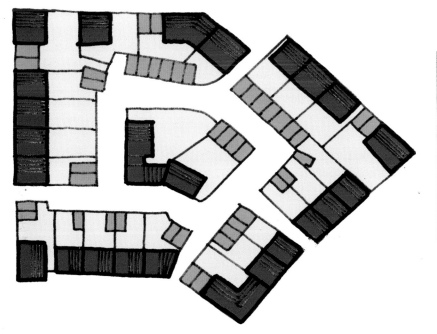

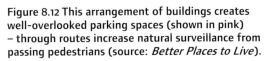

Figure 8.12 This arrangement of buildings creates well-overlooked parking spaces (shown in pink) – through routes increase natural surveillance from passing pedestrians (source: *Better Places to Live*).

Figure 8.13 This well-overlooked parking court at Bishop's Mead, Chelmsford, is obviously in the private realm (source: *Safer Places*[14]).

8.3.31   *Better Places to Live* notes that courtyard parking can be a useful addition to spaces in front of dwellings, and that courtyards which work well exhibit three main characteristics:

- they are not car parks, but places which have parking in them;
- they are overlooked by adjoining houses, or by buildings entered from the parking area (Figs 8.12 and 8.13); and
- they normally include, at most, 10 parking spaces – if there are more spaces, the courtyard layout should be broken up.

8.3.32   *Better Places to Live* also acknowledges the success of developments which depend on basement or undercroft parking, without which they would not be viable. The advantage of putting cars underground is that it preserves the street frontage, uses land more efficiently and may be more convenient for drivers accessing the building, particularly in adverse weather. However, as with courtyard parking, much depends on the location and design of the entrance.

8.3.33   *The Urban Design Compendium*[15] advises that vehicles should not be allowed to dominate spaces, or to inconvenience pedestrians and cyclists; and that a careful balance has to be struck between the desire of car owners to park as near to their dwellings as possible and the need to maintain the character of the overall setting. Parking within the front curtilage should generally be avoided as it breaks up the frontage and restricts informal surveillance. Where cars are parked in courts or squares, the design should ensure that they are overlooked by adjoining buildings.

8.3.34   *Car Parking: What Works Where*[16] provides a comprehensive toolkit for designers that gives useful advice on the most appropriate forms of car parking relevant to different types of residential development. The guidance includes examples of:

- parking in structures such as multi-storey and underground car parks;
- parking in front and rear courts;
- on-street parking in central reservations, along kerbs and at different angles, and in parking squares; and
- parking on driveways, in garages and car ports, and in individual rear courts.

14 ODPM and Home Office (2004) *Safer Places: The Planning System and Crime Prevention*. London: Thomas Telford Ltd.
15 Llewelyn Davies (2000) *The Urban Design Compendium*. London: English Partnerships and The Housing Corporation.
16 English Partnerships and Design for Homes (2006) *Car Parking: What Works Where*. London: English Partnerships.

8.3.35   The guidance includes detailed case studies that illustrate the application of these parking solutions for different locations and types of housing.

8.3.36   When drawing up parking policies or designing for new car-parking arrangements, it is recommended that local authorities and applicants seeking planning permission have regard to the good practice set out in the above guidance (and also see box). Consideration should also be given to the Safer Parking Scheme initiative of the Association of Chief Police Officers (ACPO),[17] aimed at reducing crime and the fear of crime in parking areas.

Car parking arrangements: good practice

It is recommended that the following key principles (based on *Car Parking: What Works Where*) should be followed when considering the design and location of car parking:

- the design quality of the street is paramount;
- there is no single best solution to providing car parking – a combination of on-plot, off-plot and on-street will often be appropriate;
- the street can provide a very good car park – on-street parking is efficient, understandable and can increase vitality and safety;
- parking within a block is recommended only after parking at the front and on-street has been fully considered – rear courtyards should support on-street parking, not replace it;
- car parking needs to be designed with security in mind – advice on this issue is contained in Safer Places. See also the Safer Parking Scheme initiative of ACPO; and
- consideration needs to be given to parking for visitors and disabled people.

Efficiency of parking provision

8.3.37   A key objective of PPS3 is to ensure that land is used efficiently, and to this end the total space taken up by parking needs to be minimised (*Wales:* refer to TAN 18). The more flexible the use of parking spaces, the more efficient the use of space.

Table 8.2 Efficiency of different types of parking

Level of efficiency/ flexibility	Type of parking	Comments
High	On-street	Most efficient, as parking spaces are shared and the street provides the means of access
	Off-street communal	Requires additional access and circulation space
	Off-street allocated spaces but grouped	Although less flexible in operation, this arrangement allows for future changes in allocation
	Off-street allocated garages away from dwellings	Inflexible, and largely precludes sharing spaces. Also security concerns
Low	Within individual dwelling curtilage	Requires more space due to the need for driveways, but more secure

8.3.38   Each type of solution has different levels of efficiency and flexibility (see Table 8.2).

17 See www.britishparking.co.uk.

Manual for Streets

Garages

8.3.39   Garages are not always used for car parking, and this can create additional demand for on-street parking.

8.3.40   Research shows that, in some developments, less than half the garages are used for parking cars, and that many are used primarily as storage or have been converted to living accommodation (see box).

Use of garages for parking
Recent surveys found the following proportions of garages were used for parking:
- 44% at various sites in England[18]
- 36% at Waterside Park, Kent;[19] and
- 45% at various sites in Oxfordshire.[20,21]

8.3.41   In determining what counts as parking and what does not, it is recommended that the following is taken into account:
- car ports are unlikely to be used for storage and should therefore count towards parking provision; and
- whether garages count fully will need to be decided on a scheme-by-scheme basis. This will depend on factors such as:
  - the availability of other spaces, including on-street parking – where this is limited, residents are more likely to park in their garages;
  - the availability of separate cycle parking and general storage capacity – garages are often used for storing bicycles and other household items; and
  - the size of the garage – larger garages can be used for both storage and car parking, and many authorities now recommend a minimum size of 6 m by 3 m.

Footway parking

8.3.42   Footway parking (also called pavement parking) causes hazards and inconvenience to pedestrians. It creates particular difficulties for blind or partially-sighted people, disabled people and older people, or those with

Figure 8.14 Footway parking at Beaulieu Park, Chelmsford.

prams or pushchairs (Fig. 8.14). It is therefore recommended that footway parking be prevented through the design of the street.

8.3.43   Footway parking may also cause damage to the kerb, the footway and the services underneath. Repairing such damage can be costly and local authorities may face claims for compensation for injuries received resulting from damaged or defective footways.

8.3.44   In London footway parking is prohibited, unless expressly permitted by an order. Outside London footway parking is not generally prohibited, but local authorities can prohibit footway parking through a TRO. Any such order would, however, need to be enforced, which may be costly without an awareness-raising campaign. Local authorities should therefore aim to encourage drivers to regard the footway as reserved for pedestrians, and public information and education programmes can help to influence attitudes in line with this objective.

8.3.45   It is also possible to deter footway parking through physical measures, such as by installing bollards, raised planters or other street furniture, and by clearly indicating where people should park.

18   WSP (2004). Car Parking Standards and Sustainable Residential Environments – research carried out for ODPM.
19   Scott Wilson – Surveys of garage use at Ingress Park and Waterstone Park, Dartford, Kent.
20   Some 63% of residents in Oxfordshire who did not use their garage for parking said that this was because it was used for storage, including cycle storage.
21   WSP and Phil Jones Associates (2006) unpublished reasearch.

**Derby City Council –
tackling pavement parking**

In a number of pavement parking hot-spots in Derby, the Council placed Parking on Pavements leaflets on vehicles parked on the pavement (Fig. 8.15). These leaflets give a clear message as to the negative effects of pavement parking, along with an indication of the penalties that pavement parkers could incur. Since 2002, over 300 Parking on Pavements leaflets have been placed on vehicles in hot spots, and the effect on pavement parking has been positive.

Figure 8.15 DCC's Parking on Pavements leaflets.

Dimensions for car-parking spaces and manoeuvring areas

8.3.46 Further guidance on deterring footway parking is contained in Traffic Advisory Leaflet 04/93.[22] The Department for Transport has also drawn together examples of authorities that have tackled footway parking (also see 'Derby City Council case study box').

8.3.47 Where there is a shared surface (Fig. 8.16), conventional footways are dispensed with, so, technically, footway parking does not arise. However, inconsiderate parking can still be a problem (Fig. 8.17). Parking spaces within shared surface areas which are clearly indicated – for example by a change in materials – will let people know where they should park. Street furniture and planting, including trees, can also be used to constrain or direct parking.

8.3.48 For parking parallel to the street, each vehicle will typically need an area of about 2 m wide and 6 m long.

8.3.49 For echelon or perpendicular parking, individual bays will need to be indicated or marked. Bays will need to enclose a rectangular area about 2.4 m wide and a minimum of 4.2 m long. Echelon bays should be arranged so that drivers are encouraged to reverse into them. This is safer than reversing out, when visibility might be restricted by adjacent parked vehicles.

Figure 8.16 Clearly indicated parking spaces on a shared surface in Morice Town Home Zone, Plymouth.

Figure 8.17 Untidy and inconsiderate parking.

22 Department for Transport (1993) *Traffic Advisory Leaflet 04/93 – Pavement Parking*. London: Department for Transport.

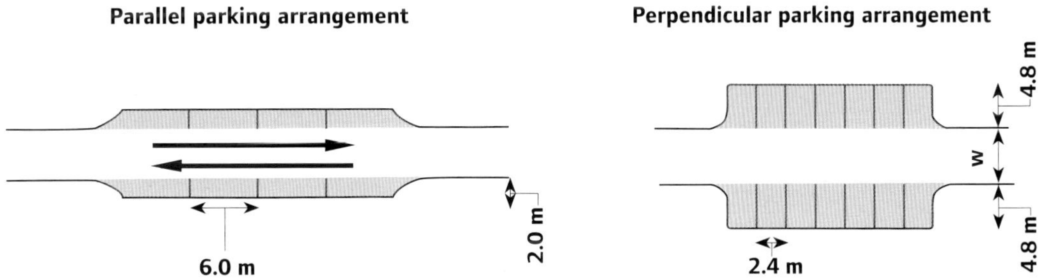

Parallel parking arrangement   **Perpendicular parking arrangement**

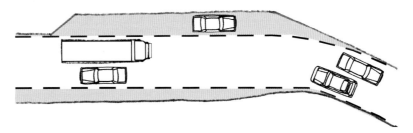

Figure 8.18 Suggested parallel and perpendicular parking arrangements.

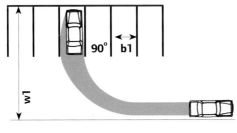

Figure 8.19 Gradual widening of the carriageway to create on-street spaces, with running carriageway checked using vehicle tracking.

8.3.50   Figures 8.18 and 8.19 show some suggested arrangements.

8.3.51   The width (W in Fig. 8.18) needed to access echelon or perpendicular spaces conveniently, depends on the width of the bay and the angle of approach. For a 2.4 m wide bay, these values are typically:

- at 90 degrees, W = 6.0 m;
- at 60 degrees, W = 4.2 m; and
- at 45 degrees, W = 3.6 m.

8.3.52   These width requirements can be reduced if the spaces are made wider. Swept-path analysis can be used to assess the effect of oversized spaces on reducing the need for manoeuvring space (Fig 8.20).

8.3.53   Where space is limited it may not be possible to provide for vehicles to get into the spaces in one movement. Some back and fore manoeuvring may be required. This is likely to be acceptable where traffic volumes and speeds are low.

8.3.54   The dimensions given above for parking spaces and manoeuvring areas can also be applied to the design of underground and multi-storey car parks. For detailed guidance on the design of these types of parking, reference can be made to guidelines prepared by the Institution of Structural Engineers (IStructE).[23]

Tracking assessment

23 IStructE (2002) *Design Recommendations for Multi-storey and Underground Car Parks.* London: IStructE.

b1 < b2
w1 > w2

Figure 8.20 The effect on overall street width requirements when wider car parking spaces are provided.

Parking spaces for disabled people

8.3.55   Detailed design specifications for parking spaces for disabled people are set out in *Traffic Advisory Leaflet 05/95*[24] and in *Inclusive Mobility*.[25] Further advice is available in BS 8300: 2001.[26] However, it is important to note that the diagrams on page 58 of *Inclusive Mobility* do not show the correct way to mark nor do they show the full range of dimensions for on-street bays for disabled people. The diagrams also show some of the kerb-mounted sign posts poorly positioned for people wishing to access their cars. Traffic signs and road markings for on-street bays reserved for disabled badge holders should comply with TSRGD and further guidance is provided in *Traffic Signs Manual Chapter 3*[27] and *Traffic Signs Manual Chapter 5*.[28]

8.3.56   It is recommended that parking bays for disabled people are designed so that drivers and passengers, either of whom may be disabled, can get in and out of the car easily. They should allow wheelchairs users to gain access from the side and the rear. The bays should be large enough to protect people from moving traffic when they cannot get in or out of their car on the footway side.

8.3.57   *Inclusive Mobility* recommends that dropped kerbs with tactile paving are provided adjacent to car-parking spaces to ensure that wheelchair users can access footways from the carriageway. (*Wales:* Further guidance on car parking standards and design for inclusive mobility will be produced in association with Welsh guidance on Design and Access Statements during 2007.)

8.3.58   The recommended dimensions of off-street parking bays are that they are laid out as a rectangle at least 4.8 m long by 2.4 m wide for the vehicle, along with additional space as set out in *Inclusive Mobility*.

8.4   Motorcycle parking

8.4.1   In 2003 there were 1.52 million motorcycles in use – representing around 5% of all motor vehicles. The need for parking provision for motorcycles is recognised in PPG13, which advises that, in developing and implementing policies on parking, local authorities should consider appropriate provision for motorcycle parking.

8.4.2   Guidance on motorcycle parking is contained in Traffic Advisory Leaflet 02/02.[29] General advice on designing highways to meet the need of motorcycles is given in the Institute of Highway Engineers (IHIE) Guidelines for Motorcycling, published in 2005.[30] Some of the guidance contained in that document has been repeated here for ease of reference.

8.4.3   The IHIE guidelines provide considerable detail on the provision of public motorcycle parking at locations such as educational establishments and workplaces, at shopping/entertainment areas and within residential areas lacking private parking opportunities.

8.4.4   Motorcyclists prefer to park close to their destination, in places where they can secure their machine. Designated motorcycle parking facilities that fail to meet these requirements will probably be overlooked in favour of informal spaces that are considered more suitable by owners.

8.4.5   Motorcycles are prone to theft, as they can be readily lifted into another vehicle. Security should therefore be a key consideration for those providing parking facilities for motorcycles.

24 Department for Transport (2005) *Traffic Advisory Leaflet 05/05 – Parking for Disabled People*. London: Department for Transport.

25 Department for Transport (2005) *Inclusive Mobility: A Guide to Best Practice on Access to Pedestrian and Transport Infrastructure*. London: Department for Transport.

26 British Standards Institute (BSI) (2001) BS 8300: 2001 *Design of Buildings and their Approaches to Meet the Needs of Disabled People*. London: BSI.

27 Department for Transport (1986) *Traffic Signs Manual Chapter 3: Regulatory Signs*. London: HMSO.

28 Department for Transport (2003) *Traffic Signs Manual Chapter 5: Road Markings*. London: TSO.

29 Department for Transport (2002) *Traffic Advisory Leaflet 02/02 – Motorcycle Parking*. London: Department for Transport.

30 IHIE (2005) *Guidelines for Motorcycling: Improving Safety through Engineering and Integration*. London: IHIE.

8.4.6   In planning for private residential parking, in most situations motorcycles will be able to use car parking spaces, but in some situations it will be appropriate to provide designated motorcycle parking areas, particularly:

·   where there is a high density of development and where car parking is likely to be intensively used; and

·   where demand for motorcycle parking is expected to be significant.

8.4.7   Where designated parking is provided, covered spaces will provide protection from the elements.

8.4.8   Physical security need not be difficult or expensive to provide. Fixed features, such as rails, hoops or posts designed to provide a simple locking point to secure a motorcycle should be considered. Where motorcycles are parked in bays with one wheel against the kerb, a simple continuous steel rail satisfies most situations (Fig. 8.21). The rail should be set at around 600 mm high to accommodate the range

of wheel sizes in use. The addition of guard railing prevents the locking rail from becoming a tripping hazard.

8.4.3   To estimate the space required for parking motorcycles, it is recommended that a 2.0 m by 0.8 m footprint is allowed per motorcycle. It is not necessary or desirable to mark individual bays. For regulated on-street parking, supported by a TRO, diagram 1028.4 of TSRGD should be used.

Figure 8.21 Secure motorcycle parking.

UFFORD

AD 1985

Upper Ufford   1⁄2
Ipswich   03.

Loudham   1^{1}2
Campsea Ash   4

Bromeswell   1^{3}4
Eyke   1^{3}4

Chapter aims

- Discuss the influence of signs on making streets successful.

- Raise awareness of the visual impact of excessive signing.

- Direct practitioners to detailed guidance.

- Examine the flexibility allowed by the Traffic Signs Regulations and General Directions 2002 and the *Traffic Signs Manual* to ensure that signing is appropriate to the street and its intended uses.

- Encourage designers to optimise signing.

9.1   Traffic signs

9.1.1   The Traffic Signs Regulations and General Directions 2002[1] (TSRGD) is a regulatory document which details every traffic sign prescribed for use in the UK. It includes all of the prescribed road markings, as a road marking is legally a sign. TSRGD also stipulates the conditions under which each sign may be used.

9.1.2   Further advice on the use of signs is contained in the *Traffic Signs Manual*,[2] which gives advice on the application of traffic signs in common situations. Chapters likely to be of particular relevance to street design include:

- *Chapter 1 – Introduction*:[3] sets out the background to, and principles of, signing;
- *Chapter 3 – Regulatory Signs*:[4] gives advice on the use of signs which give effect to traffic regulation orders (TROs);
- *Chapter 4 – Warning Signs*:[5] gives advice on signs used to warn of potential hazards;
- *Chapter 5 – Road Markings*:[6] gives advice on the use of road markings in common situations.

9.1.3   It is important that designers refer to the Traffic Signs Manual before embarking on the design of signing.

9.1.4   Supplementary advice is also published by the Department for Transport in Local Transport Notes (the LTN series) and Traffic Advisory Leaflets (the TAL series). The publications relevant to signing include LTN 1/94 *The Design and Use of Directional Informatory Signs*[7] and TAL 06/05 *Traditional Direction Signs*.[8]

9.1.5   Designers need to understand the status of these documents. Compliance with TSRGD is mandatory. *The Traffic Signs Manual*, the LTNs and the TALs are guidance.

9.1.6   On occasion designers may find that there is no prescribed sign which suits their purpose. If so, they can apply to the Department for Transport or the Welsh Assembly Government, as appropriate, for authorisation to use a non-prescribed sign. However, they should check carefully beforehand to make sure that the situation they wish to address is not already covered by TSRGD – some applications for non-prescribed signs turn out to be unnecessary for this reason.

9.1.7   Some streets feature few, or no, signs or markings. This may be appropriate in lightly-trafficked environments. It reduces sign clutter and the relative lack of signing may encourage lower vehicle speeds. However, it is worth monitoring such arrangements to confirm that the level of signing is correct.

9.1.8   In residential areas, minimal signing can work well if traffic volume and speed are low. Some designers have experimented with this approach on more heavily-trafficked streets, but there is insufficient evidence to date to be able to offer firm guidance here.

9.1.9   When planning how to sign a street, designers should note the following:
- the size of a sign should suit the speed of the traffic regardless of its purpose. It is not appropriate to use smaller signs simply because the sign is informative rather than a warning or regulatory sign. If the sign is necessary, motorists need to be able to read it;
- signs are most effective when not used to excess. Designers should ensure that each sign is necessary – they should use the flexibility within the TSRGD and associated guidance documents to ensure that signs are provided as required, but do not dominate the visual appearance of streets;

1   Statutory Instrument 2002 No. 3113, The Traffic Signs Regulations and General Directions 2002. London: TSO.

2   Department for Transport (various) *The Traffic Signs Manual*. London: TSO and HMSO.

3   Department for Transport (2004) *Traffic Signs Manual Chapter 1: Introduction*. London: TSO.

4   Department for Transport (1987) *Traffic Signs Manual Chapter 3: Regulatory Signs*. London: HMSO.

5   Department for Transport (2004) *Traffic Signs Manual Chapter 4: Warning Signs*. London: TSO.

6   Department for Transport (2003) *Traffic Signs Manual Chapter 5: Road Markings*. London: TSO.

7   Department for Transport (1994) *Local Transport Note 1/94 - The Design and Use of Directional Informatory Signs*. London: HMSO.

8   Department for Transport (2005) *Traffic Advisory Leaflet 06/05 - Traditional Direction Signs*. London: Department for Transport.

Figure 9.1 (a) Sign clutter in residential areas; (b) the yellow backing board adds clutter and its use with the flashing amber lights is counter-productive. In addition, the sign post should not protrude above the sign.

- signs which have no clear purpose should be removed to reduce clutter and to ensure that essential messages are prominent; and
- consideration should be given to incorporating colour contrast bands on poles and columns to help partially-sighted people. A single white or yellow band 150 mm deep with its lower edge between 1.5 m and 1.7 m from the ground is likely to be appropriate.

Clutter

9.1.10   Signs can clutter the street if used to excess (Fig. 9.1). Clutter is unattractive and can introduce hazards for street users.

9.1.11   Cluttering tends to take place over time by the incremental addition of signs to serve a particular purpose without regard having been given to the overall appearance of the street. It is recommended that street signs are periodically audited with a view to identifying and removing unnecessary signs.

9.1.12   In the case of new developments, some highway authorities seek to guard against having to install additional signs at their own expense later, by requiring all manner of signs to be provided by the developer at the outset. This can lead to clutter and is not recommended. The preferred way of addressing such concerns is to issue a bond to cover an agreed period, so that additional signs can be installed later at the developer's expense if required.

9.2   Designing signs

9.2.1   No signs are fundamentally required by TSRGD per se. Signs are only needed to warn or inform, or to give effect to TROs, and TSRGD simply sets out how signs must be used once it has been decided that they are necessary.

9.2.2   Designers should start from a position of having no signs, and introduce them only where they serve a clear function:

'Signs are used to control and guide traffic and to promote road safety. They should only be used where they can usefully serve these functions.'[9]

9.2.3   Street layouts, geometries and networks should aim to make the environment self-explanatory to all users. Features such as public art, planting and architectural style can assist navigation while possibly reducing the need for signs.

9.2.4   The location and design of signs and signposts should be planned to permit effective maintenance (including access for cleaning equipment) and to minimise clutter.

9.2.5   Providing additional signs may not solve a particular problem. If signs have proved ineffective, it may be more appropriate to remove them and apply other measures rather than providing additional signs. If motorists already have all the information they need, additional signing will simply clutter the environment:

'Appropriate warning signs can greatly assist road safety. To be most effective, however, they should be used sparingly.'[10]

9.2.6   The TSRGD provide significant flexibility in the application of statutory signs, including the use of smaller signs in appropriate conditions. Designers need to be familiar with the Regulations and with the published guidance, determine what conditions they are designing for and specify appropriate signs. Working drawings for most prescribed signs are available free of charge on the Department for Transport website. Designers should always start from these when adapting a prescribed sign for special authorisation.

9   Department for Transport (2004) *Traffic Signs Manual Chapter 1: Introduction.* London: TSO.
10 Department for Transport (2004) *Traffic Signs Manual Chapter 4: warning Signs.* London: TSO.

Table 9.1 Prompts for deciding on the appropriate level of signing

	Prompts
Users	• What signs are necessary to assist users, including non-motorised users? • Are directional signs needed for vehicular traffic, including pedal cyclists? • Is information provided in the necessary formats to be accessible to all? • Can navigation be assisted by means other than signs? For example, landmarks or other visual cues ,etc. • Can road markings be dispensed with in some places?
Place	• How can necessary information be integrated into the place without dominating it? • Can some pedestrian direction signs be designed to contribute to the sense of place by using a locally distinctive format? • Are traditional direction signs[12] appropriate for the setting?
Safety	• Are there any hazards that require signs? • Can significant locations, such as school entrances, health centres, local shops, etc., be indicated by a measure such as surface variation to reduce the need for signs?
Regulation	• What signing is necessary to give effect to TROs? • Is it necessary to regulate traffic or parking? • Can behaviour be influenced by means other than signing? For example, can parking be managed by the physical layout of the street?
Speed	• Are signs specified at the minimum size required for the design speed of traffic (new build) or 85th percentile speed (existing streets)? • Can traffic speeds be controlled by measures (such as planting to break-up forward visibility) to reduce the need for signs?

9.2.7   When designing for minimal signing, care should be taken that safety hazards are not left unsigned.

9.2.8   The Department for Transport may be prepared to authorise departures from TSRGD to reduce signs and road markings in environmentally sensitive streets.

9.2.9   *The Traffic Signs Manual* states that 'it is desirable to limit the number of posts in footways. Where possible signs should be attached to adjacent walls, so that they are not more than 2 metres from the edge of the carriageway, or be grouped on posts'.[11] Lighting equipment may also be mounted on walls (see Chapter 10).

9.2.10   In existing neighbourhoods, there can be legal difficulties associated with attaching signs (or lighting) to private property – this is less of a problem with new build.

9.2.11   Existing streets should be subject to a signs audit to ensure that they are not over-signed and, in particular, that old, redundant signs, such as 'New road layout ahead' have been removed.

9.2.12   The prompts in Table 9.1 will help when deciding on the appropriate level of signing for a street.

9.3   Common situations

Centre lines

9.3.1   The use of centre lines is not an absolute requirement. *The Traffic Signs Manual Chapter 5*[13] gives advice on the correct use of road markings.

9.3.2   Centre lines are often introduced to reduce risk but, on residential roads, there is little evidence to suggest that they offer any safety benefits.

9.3.3   There is some evidence that, in appropriate circumstances, the absence of white lines can encourage drivers to use lower speeds:
• research undertaken in Wiltshire found that the removal of the centre line led to a wider margin being maintained between opposing flows. There was no indication that drivers were encouraged to adopt inappropriate speeds. At 12 test sites, it resulted in slower speeds and reduced accidents, although the council had concerns regarding liability;[14] and

11 Department for Transport (2004) *Traffic Signs Manual Chapter 1: Introduction.* London: TSO. Paragraph 1.57

12 Department for Transport (2005) *Traffic Advisory Leaflet 06/05 - Traditional Direction Signs.* London: Department for Transport

13 Department for Transport (2003) *Traffic Signs Manual Chapter 5: Road Markings.* London:TSO

14 Debell, C. (2003) *White lines - study shows their absence may be a safety plus.* Traffic Engineering and Control v. 44 (9) pp316-317

- research carried out in 20 residential areas during the preparation of MfS found no relationship between white centre lines and recorded casualties (see 'Starston case study box' and Fig. 9.3).

Parking

9.3.4   In residential locations, high levels of kerbside parking and inconsiderate behaviour can create problems with access, convenience and safety. It may be necessary to manage kerbside parking through the use of restrictions indicated by signs and road markings (also see Chapter 8).

9.3.5   For designated parking spaces, markings indicating the ends of bays may be omitted if physical delineation is used, e.g. build-outs (see *Traffic Signs Manual Chapter 5*).

9.3.6   The new edition of Chapter 3 of the *Traffic Signs Manual*, which the Department for Transport expects to consult on in summer 2007, will give more guidance on footway parking and shared parking spaces.

Starston, Norfolk: effects of road markings and signs on traffic speed

Figure 9.2 Starston, Norfolk.

Starston is a village on the B1134 in Norfolk (Fig. 9.2) which was experiencing problems with excessive traffic speed. It would have required a significant number of new signs to implement a 30 mph limit. Instead, road markings were removed, signing was rationalised and natural coloured road-surfacing was used. Over half of the signs were removed and many of the remainder were replaced with smaller ones. New, locally-designed place-name signs were also installed which helped reinforce the sense of place of the village. These measures led to mean speeds being reduced by up to 7 mph.[15]

Following a Road Safety Audit, Norfolk County Council reinstalled the white lines and noted that, six months after the initial scheme opening and three months after the centre line markings were put back, there was some erosion of the earlier reduction achieved on the western approach, although they were sustained on the shorter eastern approach.[16]

The erosion of speed reduction may have been a consequence of reinstalling the white lines but drivers were also responding to other factors.

15 Wheeler, A. H., Kennedy, J. V., Davies, G. J. and Green, J. M. (2001) *Countryside Traffic Measures Group: Traffic Calming Schemes in Norfolk and Suffolk*. TRL Report 500. Crowthorne: TRL.

16 Ralph (2001) *Innovations in Rural Speed Management*. Proceedings of the DTLR Good Practice Conference. London: DTLR.

Figure 9.3 Street with no centre lining.

Figure 9.4 Kerb build-out defines parking area and provides room for planting clear of the footway.

9.3.7   Parking restrictions are often ignored where enforcement is limited. The use of planting and placing of street furniture may be a more attractive and effective way of managing parking (Fig. 9.4).

Junction priority

9.3.8   Most unsignalised junctions are designed assuming a dominant flow, with priority indicated by give-way signs and markings. There is, however, no statutory requirement for junction priority to be specified.

9.3.9   Some schemes, primarily on lower volume roads, feature unmarked junctions that require drivers to 'negotiate' their way through, with the aim of controlling speeds (Fig. 9.5). At UK residential sites studied in the preparation of MfS, unmarked junctions performed well in terms of casualties. There was, however, evidence of higher vehicle approach speeds compared with marked junctions. This may indicate an intention by drivers to slow down only when another vehicle is present. For unmarked junctions, it is recommended that the geometry on junction approaches encourages appropriate speeds.

9.3.10   Where there is a need to specify junction priority, it can be signed in three ways:
- a diagram 1003 'Give Way' marking;
- a diagram 1003 'Give Way' marking and a diagram 1023 triangle; and
- both these markings and a diagram 602 'Give Way' sign.

9.3.11   It may be appropriate to begin with the simplest option, and introduce further signing only if deemed necessary in the light of experience.

Figure 9.6 Clear and legible street name sign attached to a building.

Informatory signs

9.3.12   LTN 1/94 *The Design and Use of Directional Informatory Signs* gives guidance on directional signs for drivers. The size of lettering (defined by the x-height) should be appropriate for the traffic speed. Guidance on relating the size of signs to traffic speed is given in Appendix A of the LTN.

9.3.13   Streets need to be easy to identify. This is particularly important for people looking for a street on foot. A good system of street name plates may also make direction signs to certain sites, such as schools, churches, shopping areas, etc., unnecessary. Name plates should be provided at each junction. They should be legible with a strong tonal contrast, for example black lettering on a white background. Attaching the name plates to structures can help reduce clutter (Fig. 9.6).

9.3.14   Non-statutory signs can also contribute to the sense of place of a street. This may include examples such as village signs, as well as the permitted use of a lower panel on statutory 20 mph zone signs, which allow for scheme specific artwork and messages (Fig. 9.7).

Figure 9.7 Design contributes to sense of place and reduces clutter by incorporating several direction signs on one post.

Figure 9.5 Four-way junction with no marked priority.

10

Street furniture and street lighting

Chapter aims

· Describe how street furniture that offers amenity to pedestrians is to be encouraged, but clutter avoided.

· Comment on street furniture and lighting design relating to context.

· Explain that lighting should be planned as an integral part of the street layout.

· Recommend that where lighting is provided it should conform to European standards.

Figure 10.1 Well-designed seating.

10.1 Introduction

10.1.1 Street furniture and lighting equipment have a major impact on the appearance of a street and should be planned as part of the overall design concept. Street furniture should be integrated into the overall appearance of a street. Street audits can help determine what existing street furniture and lighting is in place, and can help designers respond to the context.

10.1.2 It is especially important that, in historic towns and conservation areas, particular attention is paid to the aesthetic quality of street furniture and lighting. Care should be taken to avoid light pollution and intrusion, particularly in rural areas. In some cases it may not be appropriate to provide lighting, for example in a new development in an unlit village.

10.1.3 Street furniture that encourages human activity can also contribute to a sense of place. The most obvious example of this is seating, or features that can act as secondary seating. In addition, street features such as play equipment may be appropriate in some locations, particularly in designated Home Zones, in order to anchor activity.

10.1.4 Where street furniture or lighting is taken out of service, it should be removed.

10.2 Street furniture

10.2.1 Excessive street furniture, including equipment owned by utilities and third parties, should be avoided.

10.2.2 Street furniture of direct benefit to street users, particularly seating, is encouraged but should be sympathetic to the design of the street and respect pedestrian desire lines (Fig. 10.1).

10.2.3 Seating is necessary to provide rest points for pedestrians, particularly those with mobility or visual impairments, and extra seating should be considered where people congregate, such as squares, local shops and schools. Guidance is given in *Inclusive Mobility*[1] and *BS 8300*[2]. Seating can sometimes attract anti-social behaviour and therefore should be located where there is good lighting and natural surveillance.

10.2.4 Although much street furniture is provided for the benefit of motorised users, it is generally located on the footway and can contribute to clutter. In some circumstances, it may be possible to reduce footway clutter by placing some of these items on build-outs.

10.2.5 Street furniture, including lighting columns and fittings, needs to be resistant to vandalism and be placed in positions that minimise risk of damage by vehicles.

10.2.6 Street furniture and lighting should be located within the limits of the adoptable highway. Street furniture should be aligned on footways, preferably at the rear edge in order to reduce clutter. Care should be taken that street furniture at the rear edge of the footway does not make adjoining properties less secure by providing climbable access to windows.

1 Department for Transport (2002) *Inclusive Mobility A Guide to Best Practice on Access to Pedestrian and Transport Infrastructure.* London: Department for Transport

2 BSI (2001) BS 8300: 2001 *Design of buildings and their approaches to meet the needs of disabled people. Code of practice.* London: BSI

Figure 10.2 Guard railing blocking pedestrian desire line - note the pedestrian in the photograph has walked around it.

10.2.7   All street furniture should be placed to allow access for street cleaning.

10.2.8   Guard railing is generally installed to restrict the movement of vulnerable road users (Fig. 10.2). In some cases guard railing has been introduced in specific response to accidents.

10.2.9   Guard railing should not be provided unless a clear need for it has been identified (Fig. 10.2). Introducing measures to reduce traffic flows and speeds may be helpful in removing the need for guard railing. In most cases, on residential streets within the scope of MfS, it is unlikely that guard railing will be required.

10.2.10   A Local Transport Note giving further guidance on guard railing is currently in preparation.

10.2.11   It may sometimes be necessary to introduce barriers to pedestrian movement. Where they are required, consideration should first be given to the use of features such as surface textures, bench seating and planting that can guide pedestrian movement whilst also contributing to the amenity of the street.

10.3   Lighting

10.3.1   Lighting can contribute to:
*   reducing risks of night-time accidents;
*   assisting in the protection of property;
*   discouraging crime and vandalism;
*   making residents and street users feel secure; and
*   enhancing the appearance of the area after dark.

10.3.2   Lighting may not be appropriate in all locations or contexts. However, if it is to be provided it should be of high quality. Lighting should generally be in accordance with BS EN 13201-2,[3] BS EN 13201-3[4] and BS EN 13201-4.[5] Guidance on lighting design is given in BS 5489-1, Code of Practice for the Design of Road Lighting,[6] to comply with the requirements of BS EN 13201.

10.3.3   Where streets are to be lit, lighting should be planned as an integral part of the design of the street layout, and in conjunction with the location and anticipated growth of planting. This may require coordination between authorities to ensure that similar standards of lighting are provided for the adopted highway and un-adopted areas, such as car parking. The potential for planting to shade out lighting through growth should be considered when deciding what to plant.

3   British Standards Institute (BSI) (2003) *BS EN 13201-2: 2003 Road Lighting – Performance Requirements.* London: BSI
4   BSI (2003) *BS EN 13201-3: 2003 Road Lighting – Calculation of Performance.* London: BSI
5   BSI (2003) *BS EN 13201-4: 2003 Road Lighting – Methods of Measuring Lighting Performance.* London BSI
6   BSI (2003) *BS 5489-1: 2003 Code of Practice for the Design of Road Lighting. Lighting of Roads and Public Amenity Areas.* London BSI

10.3.4   Lighting columns should be placed so that they do not impinge on available widths of footways in the interests of wheelchair users and people pushing prams, or pose a hazard for blind or partially-sighted people. Consideration should be given to incorporating colour contrast bands on lighting columns (see also paragraph 9.1.9).

10.3.5   Lighting should illuminate both the carriageway and the footway, including any traffic-calming features, to enable road users to see potential obstacles and each other after dark. The lighting design should ensure that shadows are avoided in streets where pedestrians may be vulnerable. Adequate lighting helps reduce crime and the fear of crime, and can encourage increased pedestrian activity.

10.3.6   While lighting fulfils a number of important purposes in residential areas, care should be taken not to over-light, which can contribute unnecessarily to light pollution, neighbourhood nuisance and energy consumption.

10.3.7   Lighting arrangements may be used to identify the functions of different streets. For example, a change of light source to provide whiter lighting can distinguish a residential or urban street from the high-pressure sodium (honey coloured) and the low-pressure sodium (orange coloured) lighting traditionally used on traffic routes. This contrast may be reduced over time if white-light sources become more commonly used in road-lighting schemes.

Lighting equipment on buildings

10.3.8   Consideration should be given to attaching lighting units to buildings to reduce street clutter (Fig. 10.3). While maintenance and access issues can arise from the installation of such features on private property, some authorities have successfully addressed these. There are likely to be fewer challenges arising from the placement of lighting on buildings in new-build streets. Where lighting units are to be attached to a building, an agreement will be required between the freeholder of the property, any existing tenants and the highway/lighting authority.

10.3.9   In attaching lighting to buildings, it should be noted that it may become subject to the Clean Neighbourhoods and Environment Act 2005.[7] It is possible that lighting could then be subject to control by Environmental Health officers if is deemed to constitute a nuisance. It is therefore important that wall-mounted lighting is carefully designed to reduce stray light.

10.3.10   Key issues in the provision of lighting in residential areas are:
· context;
· lighting intensity;
· scale; and
· colour.

Figure 10.3 Street light mounted on a building.

7   Clean Neighbourhoods and Environment Act 2005. London: TSO

Context

10.3.11 Lighting should be appropriate to the context. In some locations, such as rural villages, lighting may not have been provided elsewhere in the settlement and therefore it would be inappropriate in a new development. Often, lighting suits highway illumination requirements but is not in keeping with the street environment or the range of uses of that street. A street audit can be helpful in determining both the level of lighting and the type of equipment used in the area.

10.3.12 Over-lighting should be avoided. More detailed information is given in the *Guidance Notes for the Reduction of Obtrusive Light.*[8] This provides advice on techniques to minimise obtrusive light and recommends that planning authorities specify four environmental zones for lighting in ascending order of brightness, from National Parks and Areas of Outstanding Natural Beauty to city centres. This is helpful in determining limits of light obtrusion appropriate to the local area.

Lighting intensity

10.3.13 Guidance on the appropriate level of lighting in an area is contained in BS 5489-1 Annex B.[9] This advice provides a systematic approach to the choice of lighting class based on:
- type of road or area;
- pedestrian and cycle flow;
- presence of conflict areas;
- presence of traffic-calming features;
- crime risk; and
- ambient luminance levels.

10.3.14 BS EN 13201-2, *Road Lighting – Performance Requirements,*[10] gives details of the necessary minimum and average levels of lighting to be achieved at each of the lighting classes. For streets within the scope of the MfS, it is likely that Class ME (primarily vehicular) lighting will be inappropriate and that Classes S (for subsidiary routes) or CE (for conflict areas) should be specified.

10.3.15 Lighting levels do not have to be constant during the hours of darkness. Increasingly equipment is available which will allow street lighting to be varied or switched off based on timing or ambient light levels. This offers opportunities to design variable lighting to maximise the benefits while reducing negative impacts at times when lower lighting levels may be adequate.

10.3.16 Continuity of lighting levels is important to pedestrians. Sudden changes in lighting level can be particularly problematic for partially-sighted people.

Scale

10.3.17 As much street lighting is actually provided for highway purposes, it is often located at a height inappropriate to the cross section of the street and out of scale with pedestrian users.

10.3.18 In street design, consideration should be given to the purpose of lighting, the scale of lighting relative to human users of the street, the width of the street and the height of surrounding buildings. For example, a traffic-calming scheme in Latton in Wiltshire reduced the height of lighting columns by around 40% to make the appearance less urban. In a survey of residents, 58% thought it was a good idea, and only 3% opposed. This arrangement resulted in less intrusion of light into bedroom windows.[11]

10.3.19 Where highway and pedestrian area lighting are both required, some highway authorities installed lamp columns featuring a secondary footway light mounted at a lower height. This can assist in illuminating pedestrian areas well, particularly where footways are wide or shaded by trees. Careful design is essential to ensure that such secondary luminaries do not have a detrimental effect on the uniformity of the scheme or increase light pollution.

10.3.20 While reducing the height of lighting can make the scale more human and intimate, it will also reduce the amount of coverage from any given luminaire. It is therefore a balance between shortening columns and increasing their number.

8 Institution of Lighting Engineers (ILE) (2005) *Guidance Notes for the Reduction of Obtrusive Light.* Rugby: ILE
9 BSI (2003) *BS 5489-1: 2003 Code of Practice for the Design of Road Lighting. Lighting of Roads and Public Amenity Areas.* London: BSI
10 BSI (2003) *BS EN 13201-2: 2003 Road Lighting – Performance Requirements.* London: BSI.
11 Kennedy, J., Gorell, R., Crinson, L., Wheeler, A. and Elliott, M. (2005) *Psychological Traffic Calming.* TRL Report 641. Crowthorne: TRL.

10.3.21   Generally in a residential area, columns of 5–6 m, i.e. eaves height, are most appropriate. It should be noted that, if lighting is less than 4 m in height, it may no longer be considered highway lighting and therefore the maintenance responsibility will rest with the lighting authority rather than the highway authority.

Colour

10.3.22   The colour of lighting is another important consideration. This relates both to people's ability to discern colour under artificial light and the colour 'temperature' of the light. Light colour temperature is a consequence of the composition of the light, ranging simply from blue (cold) to red (warm).

10.3.23   In terms of discerning colour, 'colour rendering' is measured on a Colour Rendering Index of Ra0–Ra100,[12] from no colour differentiation to perfect differentiation. Generally pedestrians prefer whiter lighting. It provides better colour perception which makes it easier to discern street features, information and facial expressions. The latter can be important in allaying personal security concerns. For the lighting of residential and urban streets, an Ra of 50 is desirable – and at least Ra60 is preferable for locations of high pedestrian activity.

Other lighting considerations

10.3.24   In some contexts, lighting can contribute to the sense of place of a street, with both active and passive (reflective) lighting features blurring the boundary between function and aesthetic contribution to the streetscape.

10.3.25   As with other forms of street furniture, there are longer-term maintenance issues associated with the choice and location of lighting equipment. It is recommended that this be addressed in the planning process and that equipment which is both sympathetic to the local vernacular and for which adequate replacement and maintenance stock is available be specified.

10.3.26   In developing lighting schemes, it should be recognised that there will be an interaction between light shed and light reflected from pavement surfaces, etc. Lighting should therefore be developed in coordination with decisions about materials and other street furniture.

12 International Commission on Illumination (CIE) (1995) *Method of Measuring and Specifying Colour Rendering Properties of Light Sources*. Vienna: CIE.

11

Materials, adoption and maintenance

Andrew Cameron, WSP

11.1   Introduction

11.1.1   The quality of the environment created by new development needs to be sustained long after the last property has been occupied. This requires good design and high-quality construction, followed by good management and maintenance.

11.1.2   The latter tasks are commonly the responsibility of the local highway authority, although other public and private-sector bodies can also be involved. It is therefore important that the highway engineers responsible for adoption should be included in all key decisions from the pre-planning stage through to detailed design.

11.2   Materials and construction

11.2.1   Developers and local authorities are encouraged to consider the innovative use of materials, processes or techniques. This could be supported by local authorities adopting a wide palette of local and natural materials, bearing whole-life costs in mind.

11.2.2   The inflexible application of standard construction details and materials may not be appropriate in new housing layouts. Local authorities should be prepared to allow the use of alternative materials, landscaping treatment and features (Fig. 11.1). However, it is recommended that all materials meet the following requirements:
- easy to maintain;
- safe for purpose;
- durable;
- sustainable (including the manufacturing process and energy use); and
- appropriate to the local character.

Crest Nicholson

Figure 11.1 The use of good-quality materials achieves a sense of place without leading to excessive maintenance costs.

Figure 11.2 Good quality planting softens the street scene.

11.3 Planting

11.3.1 Planting should be integrated into street designs wherever possible. Planting, particularly street trees, helps to soften the street scene while creating visual interest, improving microclimate and providing valuable habitats for wildlife (Fig. 11.2). Care needs to be taken to preserve existing trees, particularly when changes to a street are planned (Fig. 11.3).

11.3.2 Where trees are to be used, careful consideration needs to be given to their location and how they are planted. Trench planting, irrigation pipes and urban tree soils will increase the chance of trees establishing themselves successfully, thereby minimising maintenance and replacement costs.

11.3.3 Consideration should also be given to the potential impact of planting on adjacent buildings, footway construction and buried services. Concerns have been expressed by highway authorities regarding the impact that

tree roots can have on highway drainage – this can be reduced with tree pits (see Fig. 11.4). Detailed advice on this issue is contained in *Tree Roots in the Built Environment.*[1]

11.3.4 Trees and shrubs should not obstruct pedestrian sightlines. In general, driver sightlines also need to be maintained, although vegetation can be used to limit excessive forward visibility to limit traffic speeds. Slow growing species with narrow trunks and canopies above 2 m should be considered. Vegetation should not encroach onto the carriageways or footways.

11.3.5 Maintenance arrangements for all planted areas need to be established at an early stage, as they affect the design, including the choice of species and their locations.

11.3.6 Generally, any planting intended for adoption by a public body should match standards set locally and be capable of regeneration or easy renewal if vandalised. Planting needs to be designed for minimal maintenance. Evidence that buildings and walls have been built with foundations to allow for tree growth may be required.

1 Communities and Local Government (2006) *Tree Roots in the Built Environment.* London: TSO.

Figure 11.3 Existing trees preserved in new development.

11.3.7   The planting of less robust species which require specialist skilled maintenance, or more frequent maintenance visits than usual, are unlikely to be accepted for adoption by the local or highway authority and should be avoided.

11.3.8   Alternatives to formal adoption may require innovative arrangements to secure long-term landscape management. These may include the careful design of ownership boundaries, the use of covenants, and annual service charges on new properties.

11.3.9   Funding for initial set-up costs and an endowment to generate income for maintenance (e.g. executive staff, gardening staff, site offices, equipment, machinery, stores, compost/leaf litter-bins), and community and resident facilities capable of generating regular income, may be appropriate.

11.3.10   Guidance on planting in street environments includes:
- *Roots and Routes: Guidelines on Highways Works and Trees* – consultation paper;[2]
- *Tree Roots in the Built Environment;*[3]

- BS 5837: 2005 *Trees in Relation to Construction;*[4] and
- National Joint Utilities Group (NJUG), *Guidelines for the Planning, Installation and Maintenance of Utility Services in Proximity to Trees.*[5]

11.3.11   Further advice on planting considerations is set out in Chapter 5.

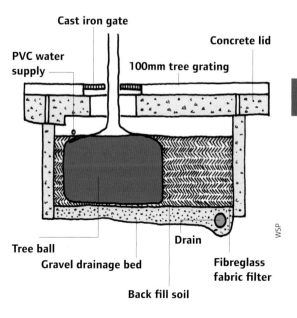

Figure 11.4 Typical tree pit detail.

2   See www.dft.gov.uk
3   Communities and Local Government (2006) *Tree Roots in the Built Environment.* London ISO.
4   British Standards Institute (BSI) (2005) *BS 5837: 2005 Trees in Relation to Construction. Recommendations.* London: BSI
5   NJUG 10 is under review at the time of writing. Please visit www.njug.co.uk/publications.htm for more details.

11.4   Drainage

Introduction

11.4.1   One of the functions of a street is to provide a route for foul water and surface water drainage (Fig. 11.5).

Foul water drainage

11.4.2   The majority of streets are designed to accommodate the disposal of foul water from buildings. This will normally take the form of drains around the curtilage of buildings which come under Part H of the Building Regulations (2000),[6] and sewers located in the street, where the relevant guidance is found within *Sewers for Adoption*.[7]

11.4.3   The adoption process for sewers is set by section 104 of the Water Industry Act 1991.[8] *Sewers for Adoption* acts as a guide to facilitate the procurement, design, maintenance and adoption of sewers, and is accompanied by a Model Agreement used by sewerage undertakers and developers.

11.4.4   An important consideration when designing sewers is their siting within the street and the impact they may have on detailed design issues. Advice on these matters is given in *Sewers for Adoption*.

Surface water drainage

11.4.5   The street provides a conduit for the storage or disposal of rainwater and, by its nature and its impact on the environment, the management of surface water runoff is a more complex matter than dealing with foul water. The Government's strategy in this area is set out in *Making Space for Water*[9], with the emphasis on the sustainable management of surface water.

11.4.6   When considering the management of surface water, designers, developers and authorities need to take account of the

6   Statutory Instrument 2000 No. 2531, The Building Regulations 2000. London: TSO.
7   Water UK (2006) Sewers for Adoption, 6th edn. Swindon: WRc plc
8   Water Industry Act 1991 London HMSO.
9   Department for Environment, Food and Rural Affairs (2005) Making Space for Water: Taking Forward a New Government Strategy for Flood and Coastal Erosion Risk Management in England. London: Defra.

Andrew Cameron, WSP

Figure 11.5 Sustainable drainage systems can form an integral and attractive part of the street.

guidance given in *Planning Policy Statement 25: Development and Flood Risk* (PPS25)[10] (*Wales*: refer to TAN 15: Development and Flood Risk[11]).

11.4.7   The planning and management of surface water discharge from buildings and highways requires a co-ordinated approach to evaluating flood risk and developing an integrated urban drainage strategy.

11.4.8   A Flood Risk Assessment (FRA) will demonstrate how flood risk from all sources of flooding to the development itself and flood risk to others will be managed now and taking climate change into account. FRA is required for planning applications where flood risk is an issue, depending on their location and size, as set out in Annex D of PPS25.

11.4.9   The responsibility for undertaking an FRA rests with the developer. However, PPS25 advocates a partnership approach, consulting with the relevant stakeholders to compile the FRA. This will involve the planning authority, the Environment Agency and sewerage undertakers. (*Wales*: refer to TAN 15.)

11.4.10   A Practice Guide[12] has been published as a 'Living Draft' to accompany PPS25. It contains guidance in the management of surface water and FRAs.The Practice Guide also covers other areas of flood risk which may be worth considering in the way streets can be used to accommodate or eliminate flood risk.

Sustainable drainage systems

11.4.11   The term Sustainable Drainage Systems (SUDS) covers the whole range of sustainable approaches to surface water drainage management. SUDS aim to mimic natural drainage processes and remove pollutants from urban run-off at source. SUDS comprise a wide range of techniques, including green roofs, permeable paving, rainwater harvesting, swales, detention basins, ponds and wetlands. To realise the greatest improvement in water quality and flood risk management, these components should be used in combination, sometimes referred to as the SUDS Management Train.

11.4.12   SUDS are more sustainable than conventional drainage methods because they:
- manage runoff flow rates, using infiltration and the retention of storm water;
- protect or enhance the water quality;
- are sympathetic to the environmental setting and the needs of the local community;
- provide a habitat for wildlife in urban watercourses; and
- encourage natural groundwater recharge (where appropriate).

They do this by:
- dealing with runoff close to where the rain falls;
- managing potential pollution at its source; and
- protecting water resources from pollution created by accidental spills or other sources.

11.4.13   The use of SUDS is seen as a primary objective by the Government and should be applied wherever practical and technically feasible.

11.4.14   Detailed guidance on SUDS is contained in the *Interim Code of Practice for Sustainable Urban Drainage Systems*,[13] Part H of the Building Regulations and Sewers for Adoption. All stakeholders need to be aware of the importance of the application of SUDS as part of an integrated urban drainage strategy for a development.

11.5   Utilities

11.5.1   Most residential streets provide routes for statutory undertakers and other services. Detailed advice on providing for utilities in new developments can be found in NJUG Guidance.[14]

11.5.2   It is best to liaise with the utility companies when the layouts of the buildings and streets are being designed. In nearly all cases this should be prior to making the planning application. Where streets are to be adopted, it will be necessary to ensure that all legal documentation required by the utility companies is completed as soon as is possible.

10 Communities and Local Government (2006) Planning Policy Statement 25: Development and Flood Risk. London: TSO.
11 Welsh Assembly Government (2004) Technical Advice Note 15: Development and Flood Risk. Cardiff: NAfW.
12 Communities and Local Government (2007) *Development and Flood Risk: A Practice Guide Companion to PPS25 'Living Draft'*. Available online only from www.communities.gov.uk
13 National SUDS Working Group (2004) Interim Code of Practice for Sustainable Urban Drainage Systems. London: Construction Industry Research and Information Association (CIRIA). See www.ciria.org/suds/pdf/nswg_icop_for_suds_0704.pdf for downloadable PDF.
14 Available from www.njug.co.uk

11.5.3   Similar principles apply to streets that are to remain private. It is important that the rights of access to the development by utility companies are set out in the management company's obligations. Residents will need to be made aware of these rights.

11.5.4   The availability and location of existing services should be identified at the outset. The requirements for new apparatus should be taken into account in the layout and design of the streets, and a balance should be struck between the requirements of the utility companies and other objectives. The locations of any existing trees or shrubs, and proposals for new planting, will require special consideration.

11.5.5   Where possible, all utility apparatus should be laid in 'corridors' throughout the site. This will facilitate the installation of the services and any future connections as the development proceeds. Consideration should be given to the use of trenches and ducts to facilitate this.

11.5.6   In designing for utilities, there are advantages in developing streets along reasonably straight lines rather than introducing gratuitous bends and curves (but see Section 7.4 regarding the control of traffic speeds). This practice will assist in simplifying utility runs, with a corresponding improvement in the efficient use of land and a reduced need for inspection chambers.

11.5.7   It may be possible to install utilities' apparatus in adopted service strips in privately-maintained land adjacent to the carriageway, provided early discussions are held with service providers and the highway authority, and that adequate safeguards are provided. Conveyance documents must incorporate perpetual rights for service providers within the service strip. Such service strips should be clearly marked and residents should be made aware of restrictions that apply to the use of these areas.

11.5.8   There have been problems with service strips where residents have not been aware of them. In addition, service strips can be unsightly and limit opportunities for planting. As an alternative, placing apparatus in the highway may be acceptable on well-connected networks, as traffic can be routed around a point closure if it is necessary to excavate the carriageway for maintenance.

11.5.9   In shared surface areas, such as in some Home Zones, the routing of services will require careful consultation between designers, utility companies and the highway authority. This consultation should take place at an early stage in the planning and design process. It may be necessary to route services in the vehicle track in some places, but as noted above this may not be a significant problem on well-connected networks.

11.6   Arrangements for future maintenance

11.6.1   It is important that the future maintenance arrangements of the streets and public spaces in a development are decided early in the design process. If the streets are to be adopted by the local highway authority, the layout and material choices need to be acceptable to the authority.

11.6.2   It is possible for streets to remain private but a properly-constituted body with defined legal responsibilities will need to be established to maintain the streets to the common benefit of residents. Further guidance on management companies is given in Section 11.9.

11.6.3   A highway authority will require legal certainty that the streets are going to be properly maintained in perpetuity by these private arrangements. In the absence of this, the Advance Payments Code contained in the Highways Act 1980[15] enables highway authorities to secure funding to meet any costs of bringing new roads up to an adoptable standard.

15 Highways Act 1980. London: HMSO.

11.6.4   A highway authority may be unwilling to adopt items such as planting and street furniture (e.g. play equipment and public art) which are not considered to relate to the highway functions of the street. If there is no private management company, arrangements can be made for such features to be maintained by another public body, such as a district or town/parish council (e.g. by designating areas of public open space).

11.6.5   In these circumstances the developer must ensure that there is agreement between the county, district and town/parish councils as to:
- which authority is best able in practice to take day-to-day responsibility for each element of planting and/or non-highway-related furniture;
- the future maintenance responsibilities, obligations and liabilities arising from such planting, street furniture etc.; and
- the apportionment of these contributions among the authorities concerned in the light of the apportioned responsibilities/liabilities.

11.7   Highway adoption – legal framework

Section 38 Agreements

11.7.1   Section 38 of the Highways Act 1980 gives highway authorities the power to adopt new highways by agreement and this is the usual way of creating new highways that are maintainable at the public expense. The Act places a duty on highway authorities to maintain adopted highways at public expense under section 41.

11.7.2   Under a Section 38 Agreement, the developer is obliged to construct the streets to an agreed standard, having first secured technical approval of the designs from the highway authority. A fee is normally payable by the developer to the highway authority to cover its reasonable costs in checking the design and supervising the construction of the works.

11.7.3   The Section 38 Agreement sets out the obligations of the developer to construct the streets and to maintain them for a set period – normally 12 months. Following the satisfactory discharge of these obligations, the new streets are automatically dedicated as public highway and are maintainable at the public expense.

Advance Payments Code

11.7.4   The Advance Payments Code (section 219 to section 225 of the Highways Act 1980) provides for payments to be made to a highway authority to cover future maintenance liabilities in the absence of a Section 38 Agreement.

11.7.5   The Advance Payments Code provides a compulsory process which involves cash deposits being made by the developer to the highway authority before building works can commence. It is an offence to undertake any house building until these payments have been deposited with the highway authority. The money securing the road charges liability is used to offset the cost of the works in instances where the highway authority carries out a Private Street Works Scheme to make up streets to an acceptable standard.

11.7.6   Thus, before any construction begins, the developer will normally be required either:

- to secure the payment of the estimated cost of the highway works under the Advance Payments Code provisions as set out in section 219 of the Act; or
- to make an agreement with the highway authority under section 38 of the Act and provide a Bond of Surety.

Private streets

11.7.7   Where a developer wishes the streets to remain private, some highway authorities have entered into planning obligations with the developer under section 106 of the Town and Country Planning Act 1990,16 which requires the developer to construct the new streets to the authority's standards and to maintain them in good condition at all times.

16 Town and Country Planning Act 1990. London: HMSO.

11.7.8   Such a planning obligation enables the developer to avoid making payments under the Advance Payments Code, as the highway authority can then be satisfied that the streets will not fall into such a condition that a Private Streets Work Scheme will be needed. The planning obligation thus provides exemption to the developer from making advance payments under section 219(4)(e) of the Highways Act 1980.

What is adoptable?

11.7.9   The highway authority has considerable discretion in exercising its powers to adopt through a Section 38 Agreement under the Highways Act 1980, but there are other mechanisms contained in the Act which help to define the legal tests for adoption.

11.7.10   Although seldom used, section 37 of the Act does provide an appeal mechanism in the event of a highway authority refusing to enter into a Section 38 Agreement. Under section 37(1), a developer can give notice to the authority that he/she intends to dedicate a street as a public highway.

11.7.11   If the authority considers that the highway *'will not be of sufficient utility to the public to justify its being maintained at the public expense'*, then it will need to apply to a magistrates' court for an order to that effect.

11.7.12   A further possibility is that the authority accepts that the new highway is of sufficient utility but considers that it has not been properly constructed or maintained, or has not been used as a highway by the public during the 12-month maintenance period. On these grounds it can refuse to accept the new road. In this case the developer can appeal to a magistrates' court against the refusal, and the court may grant an order requiring the authority to adopt the road.

11.7.13   Section 37 effectively sets the statutory requirements for a new street to become a highway maintainable at the public expense. The key tests are:

- it must be of sufficient utility to the public; and
- it must be constructed (made up) in a satisfactory manner.

In addition:

- it must be kept in repair for a period of 12 months; and
- it must be used as a highway during that period.

11.7.14   There is little case law on the application of these tests, however.

11.7.15   Highway authorities have also tended to only adopt streets that serve more than a particular number of individual dwellings or more than one commercial premises. Five dwellings is often set as the lower limit, but some authorities have set figures above or below this.

11.7.16   There is no statutory basis for the lower limit on the number of dwellings justifying adoption. The use of five dwellings as a criterion may have come from the notional capacity of private service supplies (gas, water, etc.) but it is now more commonplace for utilities to lay mains in private streets.

11.7.17   It is not desirable for this number to be set too high, as this would deny residents of small infill developments the benefit of being served by an adopted street.

11.7.18   It is recommended that highway authorities set a clear local policy on this issue.

Adoption of streets on private land

11.7.19   Under some circumstances the developer may not be able to dedicate a certain area of land as highway because he does not own it. If so, the road (or footway, etc.) can be adopted using the procedures under section 228 of the Highways Act 1980.

11.7.20   On completion of the works to the satisfaction of the highway authority, and following any agreed maintenance period, notices are posted on site. These state that unless objections are received from the owner of the land, the highway in question will become maintainable at public expense one month after the date of the notice. An inspection fee is payable in the same way as for Section 38 Agreements.

Section 278 Agreements

11.7.21   A Section 278 Agreement, under the Highways Act 1980, enables improvements to be made to an adopted highway that convey special benefit to a private body – for example, the formation of a new access to a development site, or improvements to permeability and connectivity that help strengthen integration with an existing community.

11.7.22   Before entering into such an agreement, a highway authority will need to be satisfied that the agreement is of benefit to the general public. The developer will normally bear the full cost of the works, and a bond and inspection fee is also payable, as with Section 38 Agreements.

11.8   Design standards for adoption

11.8.1   The highway authority has considerable discretion in setting technical and other requirements for a new highway. Concerns have been raised over the rigid adherence to these requirements, leading to refusal to adopt new streets. This issue was explored in *Better Streets, Better Places*.[17]

11.8.2   Highway authorities are nowadays encouraged to take a more flexible approach to highway adoption in order to allow greater scope for designs that respond to their surroundings and create a sense of place. It is recognised, however, that highway authorities will need to ensure that any future maintenance liability is kept within acceptable limits.

11.8.3   One way of enabling designers to achieve local distinctiveness without causing excessive maintenance costs will be for highway authorities to develop a limited palette of special materials and street furniture. Such materials and components, and their typical application, could, for example, be set out in local design guidance and be adopted as a Supplementary Planning Document.

11.8.4   Developers should produce well-reasoned design arguments, and articulate these in a Design and Access Statement (where required), particularly if they seek the adoption of designs that differ substantially from those envisaged in a local authority's design guide or MfS. However, provided it can be demonstrated that the design will enhance the environment and the living experience of the residents, and that it will not lead to an undue increase in maintenance costs, then highway authorities should consider responding favourably.

11.8.5   Drawings should indicate which parts of the layout the developer expects to be adopted and how the adoption limits are to be differentiated on the ground. Widths and other key carriageway dimensions, and the location and dimensions of parking spaces, should also be shown, together with full details of all planting.

11.8.6   Highway authorities would be expected to adopt street layouts complying with their Design Guide which have been constructed in accordance with the highway authority's specification of works. They would normally be expected to adopt:
· residential streets, combined footways and cycle tracks;
· footways adjacent to carriageways and main footpaths serving residential areas;
· Home Zones and shared-surface streets;
· land within visibility splays at junctions and on bends;
· trees, shrubs and other features that are an integral part of vehicle speed restraints;
· any verges and planted areas adjacent to the carriageway;
· Structures, i.e. retaining walls and embankments, which support the highway or any other adoptable area;

17 ODPM (2003) Better Streets, Better Places: Delivering Sustainable Residential Environments: PPG3 and Highway Adoption. London: ODPM

- street lighting;
- gullies, gully connections and highway drains, and other highway drainage features;
- on-street parking spaces adjacent to carriageways; and
- service strips adjacent to shared surface streets.

11.9   Private management companies

11.9.1   Any unadopted communal areas will need to be managed and maintained through private arrangements. Typical areas maintained in this way include communal gardens, shared off-street car parking, shared cycle storage, communal refuse storage and composting facilities, and sustainable energy infrastructure.

11.9.2   Where a private management company is established, it is desirable for residents to have a strong input into its organisation and running in order to foster community involvement in the upkeep of the local environment.

Index

Manual for Streets